LINEBACKER II
A VIEW FROM THE ROCK

Brigadier General James R. McCarthy and
Lieutenant Colonel George B. Allison

With a new foreword by
Major General Thomas Bussiere, Commander, Eighth Air Force

NEW EDITION

Air Force Global Strike Command Office of History & Museums
Barksdale AFB, Louisiana

LINEBACKER II | **A VIEW FROM THE ROCK**

This is a New Edition of the original 1976 book published by Air University. It has been reformatted for print and e-book, with new layout, illustrations, front-matter, and index. The main text of the book has not been altered from the original.

New Edition, 2018

Cover art by Matthew C. Koser
New Edition illustrations by Zaur Eylanbekov

History & Museums Program
Air Force Global Strike Command
245 Davis Ave East
Barksdale AFB, Louisiana 71110

ISBN: 978-0-9993317-0-5 (Perfect-bound)

For sale by the Superintendent of Documents, U.S. Government Publishing Office
Internet: bookstore.gpo.gov   Phone: toll free (866) 512-1800;   DC area (202) 512-1800
Fax: (202) 512-2104 Mail: Stop IDCC, Washington, DC 20402-0001

FOREWORD TO THE 2018 EDITION
By Major General Thomas Bussiere, Commander, Eighth Air Force

In 1909, Henry H. "Hap" Arnold, the first and only Five-Star General of the Air Force, saw his first airplane in Paris. His adventurous spirit would not allow him to sit the bench while this new technology took off. In April of 1911, Hap Arnold began learning how to fly—his instructors: the Wright Brothers! A year later, while flying his Wright Model C airplane, he went into an uncontrolled spin. He was able to recover the aircraft, but this event so traumatized Hap that he didn't know if he could ever convince himself to fly again. From that moment, Hap Arnold forged an unconquerable spirit on which our Air Force would be born. Hap Arnold not only continued to fly, but he built up American airpower to the global scale we recognize today.

Thirty years following his brush with death, Brigadier General Hap Arnold was compelled to react to the attack on Pearl Harbor. The days that followed were dark, and while the nation was reeling from a brutal blow, President Roosevelt, with the help of Hap Arnold, designed a retaliation plan relying on a very capable Lt. Col. James Doolittle to carry it out. On the day of execution, Doolittle, along with his co-pilot, Lt. Dick Cole, launched their B-25 from a carrier not knowing for sure what their fate might be. Recently, when asked what it was like to execute Hap Arnold's plan to fly a bomber off a carrier, now retired Lt. Col. Cole, the last surviving Doolittle Raider, shrugged, smiled, and said, "We just did it."

This same can do attitude and unbreakable spirit has been witnessed countless times from Bomber Airmen in the last 75 years. The Vietnam conflict's Operation LINEBACKER II called on the Mighty Eighth to contest the densest air defense of its time. It has been more than 40 years since LINEBACKER II and yet the heroic actions of American Airmen on Guam, "the rock," still echo in the minds, hearts, and history of today's Bomber Airmen. To understand the blood shed, the brotherhoods forged, and the selflessness personified by the warriors on Guam is to understand a key piece of our Air Force and more specifically, our Bomber legacy. This monograph captures not just the actions taken and the decisions made, but recounts the personal stories from the Airmen who were there—the ones who did it.

LINEBACKER II | A VIEW FROM THE ROCK

LINEBACKER II has been analyzed numerous times, examining what it meant to our country, to our Air Force, and to our Airmen. To some, the operation served as our final bargaining tool, bringing the North Vietnamese back to the negotiating table. To others, it served as an undeniable proof of the vitality of airpower and as an example of the evolution of tactics, techniques, and procedures. Still, to all, this is history to be remembered, recounted, and honored because of the legacy left by the Warrior Airmen of The Mighty Eighth.

To the Bomber Airmen of today, this book serves as a record of the hopes and dreams of every Airman that put his or her life and reputation on the (flight) line. LINEBACKER II restored hope to those locked away in the Hanoi Hilton and unlocked the chains of previous operations, allowing us to do what we were made to do, take the fight to the enemy. While negotiations for peace were in a stalemate, the nation called on the Mighty Eighth and we answered in force. This is the story about the Airmen who, with an aircraft designed and built 20 years prior, flew a wildly successful campaign against a highly capable enemy. If asked how each Airman committed himself or herself fully to complete that task, I'm sure the summary of their stories would state, "We just did it." Today's B-52s are more than 60 years old and yet it is still the indomitable spirit of the Bomber Airmen, not the aging equipment, that gets the mission done. For them, this book is not just bomber history—it is heritage.

INTRODUCTION TO THE 2018 EDITION

This is a new edition of Brigadier General James McCarthy and Lieutenant Colonel George Allison's 1976 monograph, *Linebacker II: A View from the Rock*.

Linebacker II: A View from the Rock was the Air Force's first published official history of the 11-day bombing campaign that capped off the Vietnam War. Since it was published, it has been a target for harsh criticism. Readers should remember that this was written and published as an official Air Force account. The authors' work was officially and unofficially circumscribed by a variety of factors, including the Air Force's official position, security restrictions, and limited freedom to critically comment on higher headquarters and political leaders. Also, it is written from a commander's perspective, which will always differ dramatically from the "crew dog's" view of combat. Both perspectives, as well as the materials that have emerged from Vietnam in recent decades, are valuable for understanding the relevant history. A savvy reader of history knows that every source comes with embedded biases. As an official history, this book's biases are more visible than many.

Prior to the 1976 publication of Linebacker II: A View from the Rock, much of the official record was still classified. Air Force historians and members of the CHECO and Corona Harvest teams had spent years documenting the Vietnam War, but their histories and studies were compiled for internal use. Charles K. Hopkins, the Eighth Air Force Historian, and his two staff historians, MSgt Norman A. Kramer and SSgt Raymond L. Sifdol penned what would become the first cut of the Linebacker II story; McCarthy relied heavily on their work. Portions of that history remain classified today. Since 1979, new sources have become available to western researchers, including archives and oral histories from the Socialist Republic of Vietnam. They have expanded the narrative and answered many questions, but even after 45 years, there is abundant room for further analysis and debate.

For this New Edition, McCarthy and Allison's work has been fully reformatted for print and e-book publication, with new front matter, new scans of almost all the original photographs, and re-drawn illustrations and maps by aviation illustrator Zaur Eylanbekov. However the content of the book was not altered or updated in any way from the original 1979 publication.

Read this volume with an open mind and then continue to study the subject by exploring the publications in the expanded bibliography. Avoid a single-minded view and welcome debate.

Yancy Mailes
Director
History and Museum Programs
Air Force Global Strike Command

FOREWORD TO THE 1976 EDITION

This is a narrative drawn from the era of the Southeast Asian conflict, detailing a unique event in that lengthy struggle. The event was called LINEBACKER II, a nickname like thousands of others, used to identify an operation, project, or mission associated with military affairs. It so differed from the many others, however, in its execution and outcome, that it stands alone. For the first time in contemporary warfare, heavy jet bombers were employed in their designed role to conduct extended strategic operations against the warmaking capacity of a hostile nation.

This monograph tells part of the story of Strategic Air Command's participation in LINEBACKER II. In so doing, it addresses the efforts of a complex mixture of Air Force and sister service operations, with all services working in concert towards a common goal. Rather than develop a complete chronology or blow-by-blow account, which are matters of record in other works, the campaign is pursued more from the personal perspective.

Herein is described the impact of LINEBACKER II on those in command, plus those in operations, maintenance and support who undergirded the effort, and the crewmembers. The narrative tells how they successfully met a staggering challenge. There was no book to follow. In only eleven days of intense combat operations they wrote their own book as they supported and flew the missions. That book revealed an across-the-board ability to radically change complex procedures and tactics on short notice, and a commensurate ability of aircrew and support personnel to execute them to near perfection.

In reviewing their story we find insight as to why the nation and the military need this caliber of people, who stepped forward when the need arose, demonstrated superior leadership, determination, and resiliency, did the job, and then dispersed into the more normal patterns of life. Many have since retired or separated from active service. Yet, it is clear that the ultimate well-being of our military structure in society must hinge on the continuing presence of this breed of people. Theirs was an achievement born of great ability and courage, and deserving of great honor.

LEW ALLEN JR, General, USAF
Chief of Staff

AUTHORS' ACKNOWLEDGMENTS

In preparing this text, the authors have become indebted to numerous people and agencies. Each has reinforced the valuable lesson that there is no substitute for individual expertise and personal knowledge.

The primary source of reference materials was provided by Mr. Lloyd H. Cornett, Jr., and the staff of the A.F. Simpson Historical Research Center. Particular credit is due Ms. Judy G. Endicott and Ms. Cathy Nichols, who allowed unlimited and timely access to documents.

Complementing this material were documents obtained from the offices of the Command Historian and the Directorate of Combat Operations, Strategic Air Command. Additional records of historical value were made available by the Information Division, 43rd Strategic Wing.

Ongoing research and cross-checking of details were made possible by the good services of Mr. James Eastman, Jr., and the research staff of the A.F. Simpson Historical Research Center. Where requests fell outside their purview, Miss Kenda Wise and Miss Jane Gibish of the Air University Library met every request for historical or contemporary documentation.

Lt Col Floyd Cooper of the SAC Directorate of Combat Operations assisted the authors with comprehensive specialized briefings and personal insights on the strategic aspects of LINEBACKER II. From the same directorate, Maj Arthur J. Lindemer, a veteran of four LINEBACKER II missions, gave abundantly of his time and experience to assure the quality of the finished product.

Maj Richard M. Atchison, Defense Intelligence Agency, shared numerous points of contact and recommended courses of action with the authors as the pattern of research was developing.

Periodic updates on the status of persons declared killed in action or missing in action were provided by the Missing Persons Branch, Air Force Military Personnel Center.

The authors are grateful for the time and interest which were so generously given by others who shared in the experiences surrounding LINEBACKER II. Among them are Gen James R. Allen, Lt Gen Andrew B. Anderson, Jr., Lt Gen John P. Flynn, Lt Gen Gerald W. Johnson (Ret), Lt Gen Richard L. Lawson, Lt Gen Glen W. Martin (Ret), Lt Gen Thomas M. Ryan, Jr., Brig Gen Harry N. Cordes (Ret), Col William W. Conlee, Col Hendsley R. Conner, Capt (USN) Howard E. Rutledge, Lt Col Phillip R. Blaufuss, Maj Cregg Crosby, Maj Richard L. Parrish, Maj Rolland A. Scott, and Chaplain (Capt) Robert G. Certain.

The search for photographic documentation covered the length and breadth of the country. Official unit history photographs were invaluable. Primary supplemental assistance was given by the Directorate of Information, SAC, and Mrs. Margaret Live say of the USAF Still Photo Repository. To this were added materials from Mr. John C. Dillon, Defense Intelligence Agency Photo Repository, Mr. Lawrence C. Paszek, Office of Air Force History,

LINEBACKER II | AUTHORS' ACKNOWLEDGMENTS

Ms. Sharon K. Mills, Combat Data Information Center, Lt Col Richard W. Blatter, Combat Crew magazine, Mr. William A. Ford, Air Force Magazine, and Capt John Schmick and 1st Lt James Honea, Information Division, 19th Bombardment Wing. Additional photographs from private collections were graciously provided by Lt Col Glenn Smith, Lt Col William F. Stocker, and Capt Stephen D. Cross. In the absence of all but a few original negatives, the photographic work in this document was painstakingly copied and composed from existing photographs by Sergeants Gary Zelinski and Joel J. Johnson of the Blytheville Air Force Base Support Photo Laboratory. Similar work was also done by the Barksdale Air Force Base, Louisiana, Photo Laboratory.

The authors are indebted to the many expert critiques from the field, which helped to moderate the tone while assuring accuracy. Of particular value were the studied analyses of people within the offices of the Deputy Chiefs of Staff for Intelligence and Operations Plans, and the Command Historian, Strategic Air Command, as well as those from within the Air University faculty.

Generous and experienced guidance in developing the substance and format of the text was provided by Col Ray E. Stratton, Air University, and Maj A.J.C. Lavalle, Office of the Chief of Staff, United States Air Force.

Maps and graphic displays are the work of Mr. Tommy J. Shelton and SSgt Anthony M. Olheiser of the Blytheville Air Force Base Graphic Arts Section, with supplemental work by Mr. W. Gurvis Lawson, Cartographic Information Division, Air University Library, and Technicolor Graphic Services, Inc., Maxwell Air Force Base, Alabama.

The timely and unhesitating assistance of the men of the 97th Bombardment Wing Operations Plans Division, Blytheville Air Force Base, was directly responsible for the exceptional quality of photo and map display reproductions which were forwarded to the field for final evaluation. The map displays were even further enhanced by exhaustive technical work done by Mr. Carl Roberts, Field Printing Plant, Gunter Air Force Station, Alabama.

In the final stages of preparation, Amn Melody Bridges of the 42d Air Division devoted her full attention to transcribing a mixed bag of data into an intelligible review document.

Mrs. Mary D. Gray, secretary to the 42d Air Division Commander, spent long hours in typing and proofing the final draft, performing quality control on sentence structure and content as she progressed.

MSgt Ronald I. Wilson and the members of the Blytheville Air Force Base Reproduction Center devoted special attention to the printing and collating of the draft manuscript.

Finally the authors acknowledge typing and proofing assistance by Mrs. Dorene Sherman, Headquarters SAC Command Section, additional art work by SAC/CSP, and the reviewing assistance of several Hq SAC agencies, all in the spirit of positive assistance: DO, HO, JA, LG, OI, XO, and XP.

ABOUT THE AUTHORS

Brig Gen James R. McCarthy is the Commander, 42d Air Division (SAC). His wings comprise all of Strategic Air Command's forces in the Southeast United States and Ohio. He is a command pilot and radar navigator with more than 7000 hours flying experience. He has flown more than 1,200 combat missions in Southeast Asia in such diverse aircraft as the B-52, F-4E, KC-135, C-130, and C-123. During four and one-half years experience there he served in a wide variety of assignments, including KC-135 Squadron Commander, KC-135 Wing Commander, B-52 Wing Commander, and Consolidated Aircraft Maintenance Wing Commander. During LINEBACKER II he led his wing on two B-52 raids against Hanoi, North Vietnam. He was the Airborne Mission Commander on 26 December 1972, the largest raid of the LINEBACKER II campaign.

Lt Col George B. Allison is a master navigator with more than 4,200 hours flying time, some 2,100 of which were in the B-52. He has nine years crew experience in bombardment aircraft and seven years staff experience in planning and instructing SAC bombing and navigation operations. He flew 76 B-52D combat missions in Southeast Asia from both U-Tapao Royal Thai Navy Airfield, Thailand, and Andersen Air Force Base, Guam. More than one-half of these were as a crew radar navigator during the last three months of 1972. They included 20 missions against targets in North Vietnam, two of which were LINEBACKER II missions from Guam against Hanoi. He, along with then Colonel McCarthy and thousands of other men and women, was on the Rock when it happened.

TABLE OF CONTENTS

FOREWORD TO THE 2018 EDITION	III
INTRODUCTION TO THE 2018 EDITION	V
FOREWORD TO THE 1976 EDITION	VI
AUTHORS' ACKNOWLEDGMENTS	VII
ABOUT THE AUTHORS	IX
TABLE OF CONTENTS	XI
AUTHORS' INTRODUCTION	XIV
CHAPTER 1: PRELUDE	1
PEACE IS AT HAND	1
THEY CALLED IT LINEBACKER II	2
A MIRROR OF HISTORY	4
CHAPTER 2: THE STAGE IS SET	11
BUILDUP OF THE FORCE	11
Aircrew Training	17
The "Bicycle Works"	20
"Bag Drags"	22
Charlie Tower	23
ARC LIGHT Center	26
HIGH THREAT PRESS-ONS	28
Spare Aircraft	32
A TASTE OF THINGS TO COME	32
CHAPTER III: ACT ONE	37
THE DIE IS CAST	37
DAY ONE—HOW DO YOU LIKE THE SUSPENSE?	48
DAY TWO—REPEAT PERFORMANCE	64
DAY THREE—THE DARKEST HOUR	75

LINEBACKER II | A VIEW FROM THE ROCK

CHAPTER IV: ACT TWO	**87**
DAY FOUR—THE PLOT SHIFTS	87
DAY FIVE—WORK, COOPERATION, AND PREPARATION	96
DAY SIX—BACK TO ACTION	103
DAY SEVEN—AN ISLAND PARADISE?	108
CHAPTER V: INTERLUDE	**117**
A MOMENT OF PEACE	117
Chaplains	117
CHANGE IN THE SCRIPT	119
CHAPTER VI: ACT THREE	**127**
DAY EIGHT—ONE FOR THE RECORD BOOKS	127
Total Force Participation	140
DAY NINE—LAST MOMENT OF PAIN	149
DAY TEN—THE END IS IN SIGHT	156
Bomb Loaders	160
DAY ELEVEN—THE CURTAIN COMES DOWN	164
CHAPTER VII: POSTLUDE	**171**
BUSINESS AS USUAL	171
SUMMARY	173
ASSESSMENT	174
APPENDIX	**179**
ORGANIZATIONS AND COMMANDERS, 1963-1974	179
GLOSSARY	**185**
BIBLIOGRAPHY	**195**
EXPANDED BIBLIOGRAPHY FOR THE 2018 EDITION	199
Books	199
Periodicals	199

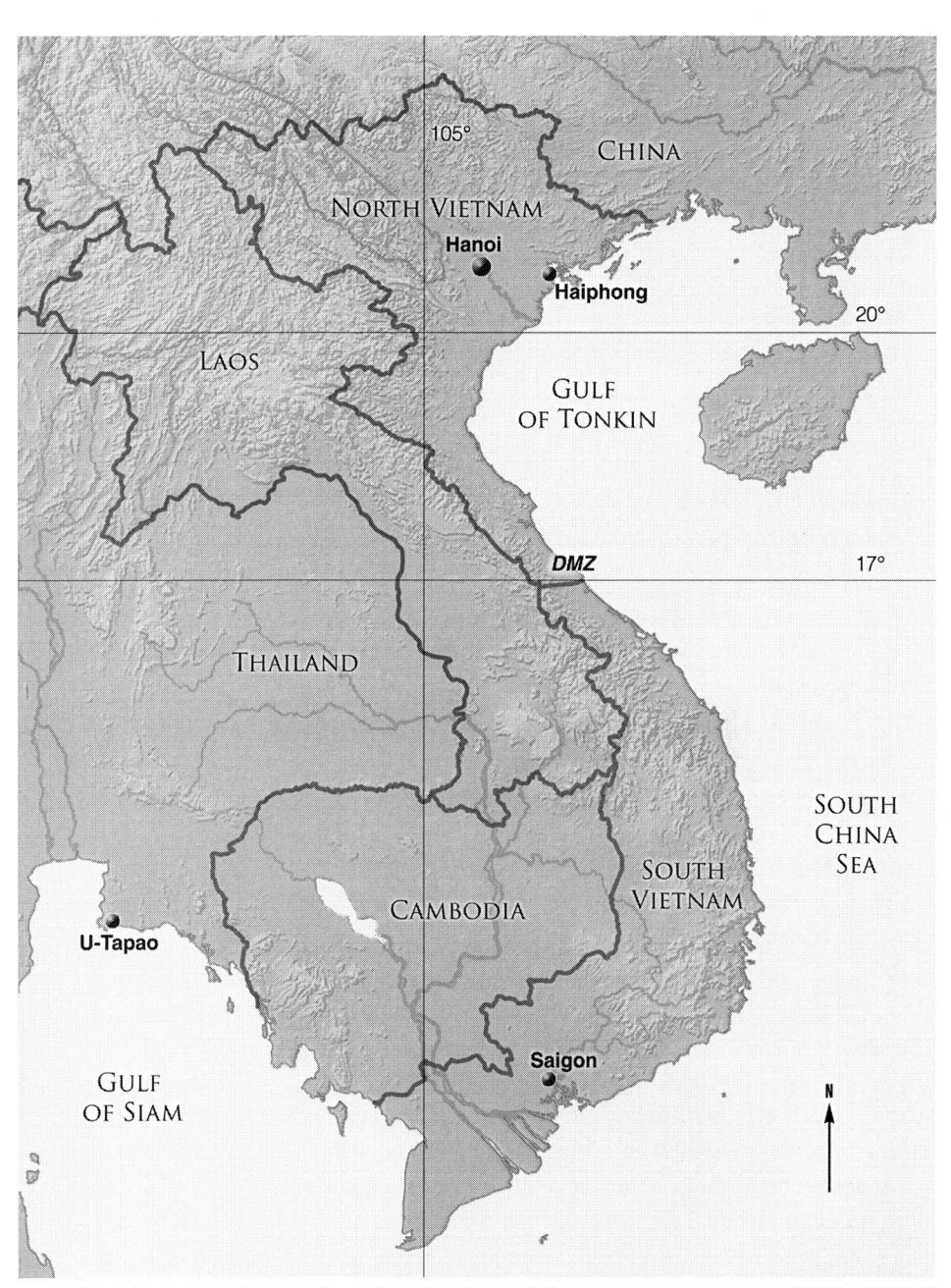

Map of Southeast Asia, showing the 17th and 20th parallels, important references in LINEBACKER II operations.

AUTHORS' INTRODUCTION

We all owe it to the heroic participants in any endeavor to recount, as best we can, the details of that portion of history which was, for them, real and immediate. So it should be for those who served with honor in Southeast Asia, regardless of confused issues, purpose, and outcome there. They were not party to that confusion, nor were they responsible for a significant lack of popular support for the conflict. Their dedicated, competent service to country was no less glorious or exhausting than that of their forebears, who have won the accolades of history. In time, the more noteworthy events of the Southeast Asian conflict may well stand with such historic epics as the Normandy Invasion, Bastogne, Midway, Iwo Jima, Pusan, and Inchon as monuments to the determination, capability, and valor of the American citizen in uniform.

It is a privilege for the authors to focus on one such event in Southeast Asia, one made monumental by the sheer scope of physical effort. Yet to be determined is the honor which may accrue to the people who translated concept into reality, and in so doing placed themselves in the forefront of that particular history.

The authors are continually aware, and the reader must ever be mindful, that the narrative developed here is only one page from a chapter of gallantry in combat. That chapter fits into a much larger book.

The symbolic page focuses on the involvement of Strategic Air Command (SAC) forces during the LINEBACKER II (Two) campaign in December 1972. More specifically, it recounts events and cites examples in support of central ideas which are drawn from the authors' first-hand experiences, complemented by the experiences of others. To that extent, our purpose is to tell of some of the outstanding performances of 12,000 men and women stationed at Andersen Air Force Base, Guam, fondly renamed "Andy" or "The Rock" by those who manned it.

"The Rock" is an appropriate name for Guam. It is actually the top of a 35,000 foot-high mountain which has its roots in the deepest part of the Pacific Ocean, the Mariana Trench. The top 1,300 feet rise abruptly out of the water and form an island 32 miles long, and a scant 11 miles at its widest point. This mountaintop, higher in its own peculiar fashion than Mount Everest, is located 3,800 miles from Hawaii and 2,900 miles from Hanoi, North Vietnam. Yet, the remote island found itself in the thick of an immediate and time-sensitive war. Disassociated from its adversary by the equivalent of an entire ocean, it was nevertheless inextricably wedded to the mainland of Southeast Asia by the tools of modern warfare.

The distance and remoteness were offset by the capabilities of our nation's venerable heavy jet intercontinental bomber, the Boeing B-52 Stratofortress, supported by fuel from the KC-

LINEBACKER II | AUTHORS' INTRODUCTION

135 air refueling tanker. A long-time veteran of the war, the B-52 had gained familiarity amongst friend and foe alike for its effectiveness in a variety of combat roles. It was because of this widespread familiarity that the giant airplane acquired an interesting nickname. At some unknown point in the lengthy war, someone referred to it as "that Big Ugly Fat Fella," and named it the "BUFF" for short. Obviously intended as a scornful label, it had much the opposite effect on those who flew as its crews, and has been regarded ever since as a name of affection and respect. If that respect was lacking from any other quarter, it ought not to have been by the war's end. The results of hundreds of long-range sorties, delivering thousands of tons of high explosives with consistent accuracy on targets hidden by monsoon weather and darkness, stand as their own proof of what the proud BUFF meant to the culmination of the war effort. While these sorties were not conducted in the heightened psychological drama of such historic events as the Schweinfurt, Ploesti, or Doolittle Raids, they may have been ultimately more decisive. But it remains for people other than those involved to say it.[1]

It is the primary intent of this work to give insight as to how the implementation and execution of so intense and extensive an operation as LINEBACKER II was made possible, particularly from such a formidable global distance.

But note that if the reader becomes preoccupied with the perception that it all began and ended on Guam, a grave disservice will have been performed. The whole chapter, and not the page, must be the recurring thought—if not in the written word, then at least in the mind's eye. The text will speak, insofar as it is pertinent to the point at hand, of the involvement of other people from other places. The reader's continued reflection on the magnitude of a great cooperative venture will help to balance the picture.

The LINEBACKER II campaign was unequivocally a team effort, on the grand scale. Tens of thousands of people put it together and, using thousands of items of war and support machinery, made it work. It is with hesitation that one even begins to mention the participants, because there at once arises the challenge of where to terminate the list. Looking at it from the broad perspective, major areas of involvement are offered.

The location with closest association was U-Tapao Royal Thai Navy Airfield, Thailand, sister base and companion in the B-52 heavy bombardment operations. Her residents truly met the test, for while their missions were significantly shorter, they "bit the bullet" more often. With less than one third of the available aircraft and crews, they flew over 45 percent of the effort, and it was not difficult to find there crewmembers who flew five or more times in the eleven-day period.

From Thailand, the Philippines, Taiwan, and Okinawa came the KC-135 aerial tankers, which not only gave the bomber force its essential flexibility, but were expected to be (and were) virtually everywhere at once, providing similar flexibility to the swarms of support aircraft.[2]

xv

LINEBACKER II | A VIEW FROM THE ROCK

The latter came from the entire theater of operations, and to categorically label them as "support" clouds the issue. Many had complex and hazardous missions of their own, done alone or in groups without collateral support.[3] The F-111, F-4, and U.S. Navy and Marine Corps aircraft tactical strikes come immediately to mind, as do the preemptive strikes against known or suspected enemy defensive positions, executed by other Thailand and carrier-based fighters and fighter-bombers.

Closely related in time to the climactic moments of each day's efforts were the activities of the F-4, EB-66, Navy and Marine EA-6, and EA-3 defensive countermeasures aircraft, preparing the way for the main force with chaff deployments and supplementing it with their electronic jamming and deception equipment. Protective F-4s flew MIG CAP (combat air patrol), while others protected their companion chaff deployers. F-105 and A-7 "Iron Hand" flights, in concert with yet more F-4s, formed Hunter/Killer teams to apply unrelenting pressure on the most serious threat to success of all—launches of the deadly SA-2 surface-to-air missile (SAM).

The search and rescue (SAR) people were there, as they always have been, often turning potential disasters into happy endings, both on land and at sea.

Inserted in the daily flow of activities were operations by four reconnaissance systems— "OLYMPIC TORCH" U-2R, RC-135M "COMBAT APPLE", DC-130 "BUFFALO HUNTER" launches of drones, and "GIANT SCALE" SR-71 missions.[4]

At ground level were all of the people who coordinated the air effort and helped keep it together. Here we refer primarily to those in the forward area control capacity,[5] but are led unavoidably to reflect on the hours and days of planning, analyzing, coordinating, evaluating, etc., which were the stock in trade of people thousands of miles away. Some, in fact, worked half a world away at SAC Headquarters. These, too, served with true dedication, sometimes in the face of caustic criticism for their lack of prophetic powers.

Inevitably, we are drawn to all of the other people who, while they didn't fly and fight, made it possible for the proportionate handful who did to carry the results of their labors to an ultimate conclusion. As a sign next to the aircrew billets at Andersen put it, "Our Mission Is BOMBS ON TARGET—It Takes Us All."

But we are getting somewhat ahead of our story. We only seek to make it clear to the point of redundancy that it was not solely a select cadre of SAC crewmembers and their B-52 bombers who flew LINEBACKER II.

Since the final outcome of the Southeast Asian conflict has denied us access to any accurate records compiled by the North Vietnamese, we are missing a major part of the story.

Fortunately, many gaps in their portion were filled in by some superlative reconnaissance and intelligence gathering efforts on our part, and by world news media. Nevertheless, the authors are acutely aware that this monograph is constructed without benefit of all the facts. The ensuing pages represent a determined effort to do justice to accuracy with such information as is available.

Finally, it is our intent to pay tribute, by selected example, to thousands of magnificent people who deserve, at the very least, a pat on the back and a heartfelt thanks. They earned far more than it is within our power to give them. We give them the best we have to offer—our pride in their greatness and our gratitude.

This work, a labor of love, is dedicated to those gallant airmen who did not return from LINEBACKER II.

The Authors
Blytheville Air Force Base, Arkansas.
21 February 1978

The ARC LIGHT Memorial B-52, Andersen Air Force Base, Guam

LINEBACKER II | A VIEW FROM THE ROCK

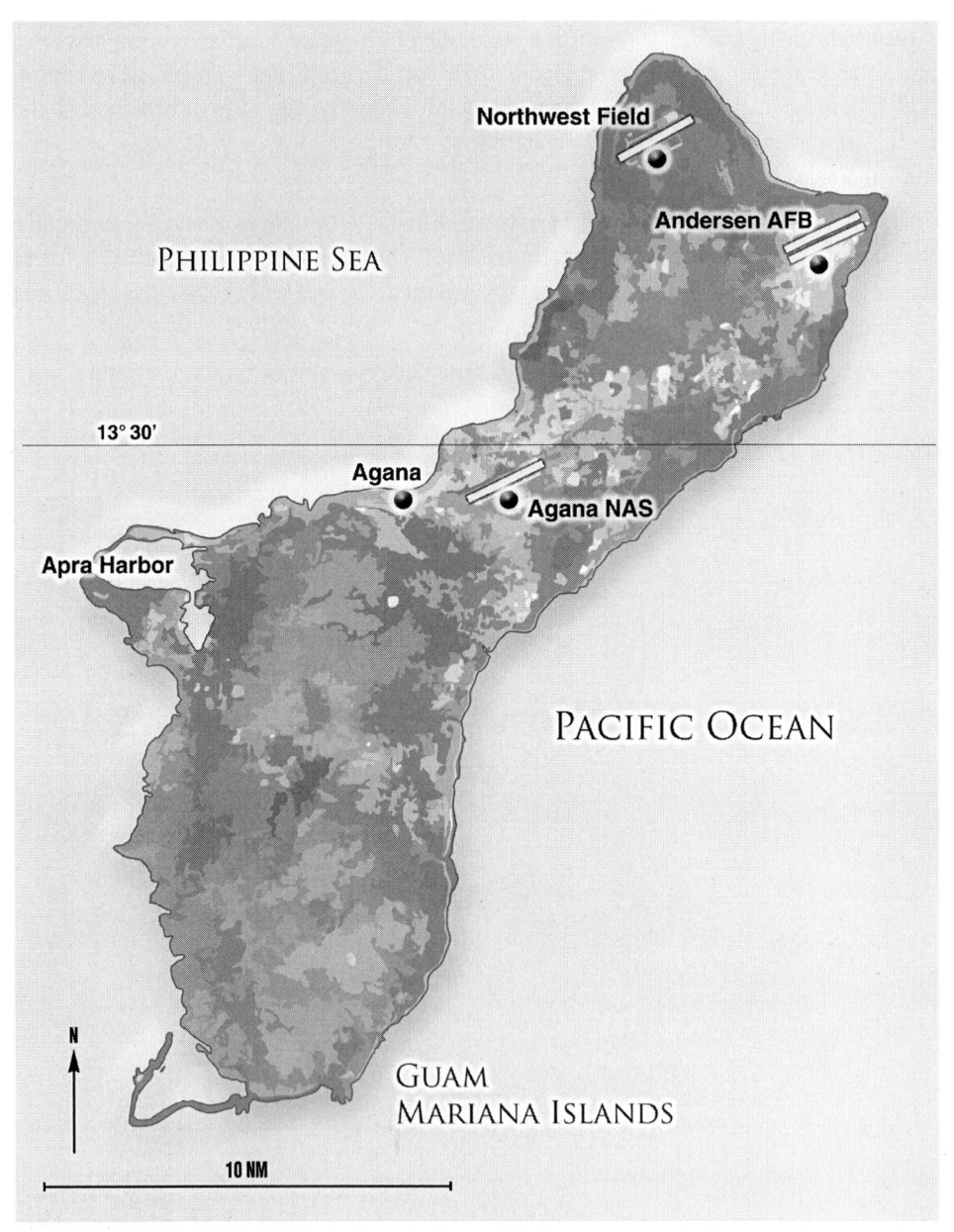

The island of Guam, Mariana Islands. B-52 LINEBACKER II operations took place at Andersen Air Force Base.

NOTES

1 W. Scott Thompson and Donaldson D. Frizzell, ed., *The Lessons of Vietnam,* New York, NY: Crane, Russak & Co., 1977, pp. 105 and 177.
2 *History of Eighth Air Force, 1 July 1972 30 June 1973,* Volume II, Narrative Part II, Andersen AFB, Guam, Mariana Islands, 23 August 1974, pp. 347-349. Hereafter cited as *8AF History, V II.* SECRET
3 *USAF Air Operations in Southeast Asia, 1 July 1972-15 August 1973,* Volume II, CORONA HARVEST, Prepared by HQ PACAF with support of SAC, Hickam AFB, HI, 7 May 1975, p. IV-243. Hereafter cited as *USAF AIROPS.* TOP SECRET
4 *SAC Participation in LINEBACKER II,* Volume I, Basic Report, HQ SAC/[XOO], Offutt AFB, NE, 5 January 1973, pp. 1-1 to 1-3. Hereafter cited as *SAC Participation.* TOP SECRET
5 *USAF AIROPS,* p. IV-312.

CHAPTER 1 | PRELUDE
PEACE IS AT HAND?

It was December 18th, 1972 and a new milestone in air power was approaching. The stage had been set for the longest ranged bombardment missions in the annals of aerial warfare.

Darkness had settled over the countryside of North Vietnam as the first aircraft in a night-long force of 129 B-52 Stratofortress heavy bombers made its turn over the Initial Point (IP) and was in-bound to the target. At 1943 (7:43 PM) Hanoi time, bombs would start impacting Hoa Lac Airfield, 15 miles west of the capital city.[1] Within the next few hours, and subsequent days, the theory and viability of high altitude strategic bombardment was going to be put to the test once more. SAC's flight crews and aircraft, while attacking some of the most heavily defended targets in the history of aerial warfare,[2] would be answering the essential question: Could high altitude bombers penetrate to and successfully attack targets defended by modern surface-to-air missiles and high performance jet fighters? We had bet a lot of irreplaceable aircraft and lives that they could. If we were wrong, then the United States would lose a significant part of its long-range bomber fleet. Along with that loss would be an incalculable loss in credibility and military stature.

We had been brought to so dire a showdown by an apparent inability to communicate via any other medium. The last argument of kings was being employed to bring home to the vacillating North Vietnamese (NVN) that our national intent was to bring an end to the conflict under the terms which had been painstakingly developed at the negotiating table. Heavy bombardment on a concentrated, massive scale against the North Vietnamese ability to make war was the method selected to bring the point home.[3]

This test of strategic capability, which systematically vindicated itself with each new day,[4] need not have come to pass. In fact, it came as a shock. Less than two months before, Presidential Advisor Dr. Henry A. Kissinger had, in full coordination with the North Vietnamese, announced that ". . . peace is at hand."[5] Considering the frustrations and complexities of the peace negotiations up to that point, his straightforward comments were packed with deep diplomatic significance. Moreover, there was a splendid aura to the phrase which captured the imagination of the world. It stuck in the minds of all peaceloving individuals, especially those doing the fighting, and actions and attitudes were constantly tempered by the hopeful thought, "Will today be the day?"

Seldom do statements of such deep implication die on the vine, and the hope which they generate endures even in the face of conflicting evidence. Among the more perceptive students of history at Andersen Air Force Base who watched the timeliness of the comments slipping away, expectations still remained high for a conclusive pronouncement from either Doctor Kissinger or President Richard M. Nixon. Therefore, as the old saying goes, it was "like being hit with both barrels" when the first warning order for the initial raids was received on December 15th. Not only were the hopes dashed, but they were replaced with the difficult realization that an unknown, but unavoidable, price was to be paid.

With that warning order—a message transmitted from the Chairman, Joint Chiefs of Staff (JCS) to the Commander-in-Chief, Strategic Air Command (CINCSAC) and relayed to the Eighth Air Force Commander[6]—LINEBACKER II became the complete occupation and preoccupation of the air forces marshalled on the Rock.

THEY CALLED IT LINEBACKER II

The selective process whereby the campaign came to be known as LINEBACKER II is, in itself, of no great consequence. It could have just as well been nicknamed anything else. As it turned out, our President was the nation's number one football fan, and the nickname was a natural outgrowth not only of that, but of the campaign which preceded it. The title LINEBACKER II presupposes a LINEBACKER I, which indeed there was.

This was a campaign involving an extensive integrated effort by all types of U.S. airborne forces to interdict the enemy logistics throughout Southeast Asia, mainly in the North.[7] For the heavy bombers, much of this emphasis was concentrated in the panhandle of North Vietnam and the Demilitarized Zone (DMZ). Implemented in the spring of 1972 by the Joint Chiefs of Staff, LINEBACKER I reflected previous efforts of past years under a nickname of historic proportions, ROLLING THUNDER.[8] One of the notable differences between ROLLING THUNDER and LINEBACKER I was that, for the first time in the Southeast Asian conflict, limited B-52 strikes were conducted against targets on the northern coast of North Vietnam.

As of 22 October, LINEBACKER I operations came to an end.[9] Then, suddenly, the need to identify a major new thrust to the war arose. Were it not for the uniqueness of the operation, where widespread interdiction was replaced by concentrated strategic bombardment, LINEBACKER II perhaps might have been just another milestone in a reinstated LINEBACKER I.[10]

CHAPTER 1 | PRELUDE

Such was not the case, and the history of warfare inherited a name which may well stand the test of time. It is somewhat ironic to note in passing that this great offensive air campaign should be executed under a nickname made famous by a defensive position in football. Thus is fate, and it was surely neither the first nor last time that such an incongruity has crept into the affairs of war.

A camouflaged Stratofortress releases its ordnance on a target in Vietnam. The concentrated strategic targeting of LINEBACKER II called for a more rapid release sequence than that shown here.

Perhaps its old forebear, ROLLING THUNDER, would have more closely captured the tone of the occasion. Assuredly, the bombardments were thunderous to the point of being described as "earthquakes" by those on the ground.[11]

Lacking within its own title the essence of its true nature, and due mainly to world press coverage and the season of the year, LINEBACKER II acquired some interesting colloquial titles. Recognizing that the official nickname would have little or no meaning to the public, the press at first labelled it the "Air Blitz." As its intensity continued unabated and then, as its aftereffects came to light, it was referred to as "the Siege of Hanoi." A bizarre nickname, born in the flush of the moment, was "the Christmas bombings," despite the fact that this was the one day when a premeditated reprieve was granted as a gesture of peace.[12]

Within military circles, "The December Raids" was popular, but the name most often heard after the campaign had concluded was "The Eleven Day War."[13] This title has stuck, much to the deserved chagrin of people who had fought through the many years preceding it. However, a name must communicate to have any lasting value. When speaking of "The Six Day War" between Israel and the Arab bloc in 1967, the dynamics of the portion in quotes carry the message.

The nickname was LINEBACKER II. "The Eleven Day War" tells the story.

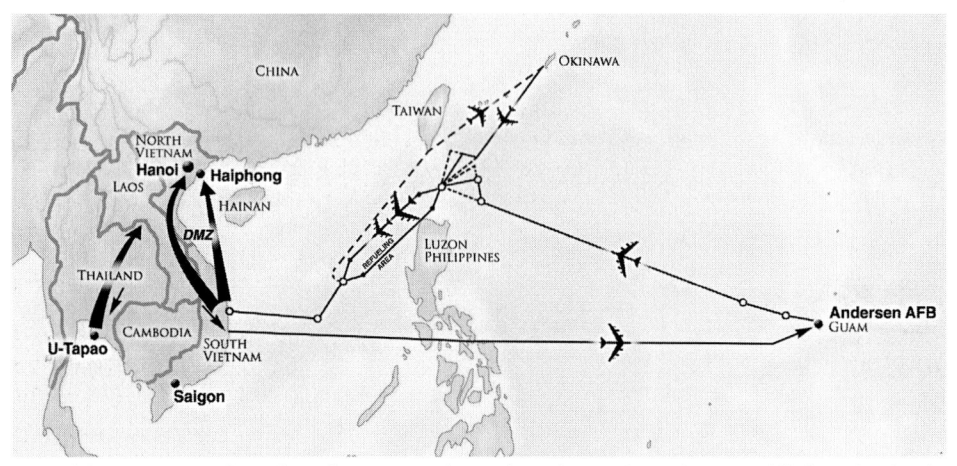

Map of the Western Pacific and Southeast Asia. The westbound route shows the general flight path taken by the B-52 forces to arrive at the mainland, from where they dispersed on a variety of routes into North Vietnam. Most sorties recovered to the South and back to Guam. Diversions from these basic routes were sometimes made to meet critical timing requirements or to perform supplemental inflight refuelings.

A MIRROR ON HISTORY

It was reminiscent of another era, recalling a time when many of LINEBACKER II's participants had not yet been born. But here it was again—history revisited. For some, the parallel was unavoidable. It was the summer of 1943 transposed into the future. Eighth Air Force was again carrying the war to the enemy's heartland. While the long-term geopolitical goal was not the same, the immediate objective was. The enemy's capability to support the war effort must be crippled by strategic bombardment. Variations in specific targeting were driven by the situation at hand, but the principle was the same. Ploesti, Regensburg, and Schweinfurt had become Hanoi and Haiphong.

The seemingly interminable lines of B-52s moving relentlessly into takeoff position recalled old photographs of a similar nature, showing desert-pink B-24s shimmering and dancing in the Libyan heat, or dull gray B-17s in the gloom of an English morning.

CHAPTER 1 | PRELUDE

Whether the men and planes launched across the English Channel, Mediterranean Sea, or Pacific Ocean, they shared a common prospect of great hazard and long hours in flight. The longest missions in the European theater had initially become the longest in Southeast Asia. By coincidence, the 14-hour Ploesti mission length was duplicated by the early LINEBACKER II raids from the Rock. Later missions were even more extensive—exceeding 18 hours in some cases.[14]

Three cells of B-52Ds line up for launch on a LINEBACKER II mission. This scene was being repeated on every available taxiway. The swayback to Andersen's runways is clearly seen.

For the B-52s flying out of Andersen Air Force Base, the distance and time of the missions represented the longest sustained strategic bombardment flights ever attempted. The direct flying distance from Guam to Hanoi is approximately 2,650 statute miles. The actual combat routing was much longer, because the heavy bombers had to fly internationally approved routes to Southeast Asia which avoided commercial air traffic. Once in the forward area, the routes were planned both to avoid as many enemy defenses as possible and to achieve precision timing. These requirements made the one-way distance to the targets exceed 3,000 miles early in the campaign.

This was not the longest routing, however. On December 26th, 1972, a portion of the fleet from Andersen flew more than 4,000 miles while the rest of the force joined up with them. It was nine and one-half hours after launch before they dropped on their targets. They then completed an approximately 8,200-mile round trip after more than 18 hours in the air.

In such circumstances, sufficient fuel becomes a key element. For the old timers in the B-24s, this took the form of extra tanks in the fuselage. The B-52Ds, old models modified for heavy bomb loads, exercised their inflight refueling capability as a matter of necessity. The newer B-52Gs, which have larger fuel cells and more efficient engines, were usually assigned the longer mission routes. This dictated inflight refueling for them as well. In any case, all launched with as much fuel as they could safely carry. As one of the World War II raiders so aptly noted, it ". . . would certainly be ironic—having just destroyed thousands of gallons at Ploesti, we ourselves should perish for want of it."[15]

Just as other "normal" activities ground to a halt for a select group of men in 1943, so an ominous and thought-provoking standdown of nearly the entire 1972 bomber force on the day before the first raids presaged things to come.16[16] It was different 29 years earlier, but the fascinating comparisons were still there. In the days prior to the Ploesti mission, the force was completely removed from the war and participated in intensive low level training missions against a mock target, duplicating the refineries, which had been constructed in the desert.

No such preparation was the case for the LINEBACKER II bomber crews. They were told the names of their targets, given only a few minutes of detailed briefing and target study, and sent on their way. Complicated routings and timing maneuvers often had to be deciphered, analyzed, and digested as time would permit enroute to the target. In spite of these obstacles and an intense combat environment, only one crew is recorded as failing to drop their bombs solely because of inability to navigate to or identify the target. In that case, it was a self-imposed withhold of release. The radar navigator (RN) was not absolutely certain of his aiming point, and the need for pinpoint accuracy was paramount.[17]

Then, there were the defenses to consider. Here we see comparisons and counterpoints. By coincidence, and due mainly to the imagination and drive of the German defense commander, Ploesti was ringed with what was probably at the time the most intense and sophisticated antiaircraft defenses in the world, certainly more so than those around Berlin. Hanoi was similarly described in 1972 terms.[18] The now-famous cliche that "the flak was so thick you could walk on it" had its modern day equivalent when Captain Hal Wilson and his co-pilot, Capt "Charlie" Brown, flying out of U-Tapao with the call sign "Rose 1" on the first day, reported over the radio shortly before being shot down that they had "wall-to-wall SAMs."[19]

CHAPTER 1 | PRELUDE

View from the cockpit of the third aircraft in a three-ship right echelon formation. This was the basic formation used for in-flight refueling as the bombers approached their tankers.

On the other side of the coin is the historical fact that, although antiaircraft artillery (AAA) took a substantial toll, the greatest aerial challenge of World War II came from the fighter defenses. While present in North Vietnam, they did not mount the degree of threat to the LINEBACKER II force which had been expected. Still, no flyer was freed from the haunting knowledge that they might indeed show up at any moment, and there were enough sporadic engagements to keep the penetrators on edge.[20] Nevertheless, over North Vietnam it was essentially an air-to-ground thrust, parried by a ground-to-air reaction.

It is further appropriate to point out, to the everlasting credit of the bomber crews in 1943, that they were on their own in the raids previously mentioned. This was tragically portrayed in the missions conducted well beyond the radius of action of friendly fighter protection. Vicious assaults were made on the Regensburg-Schweinfurt forces by swarms of German fighters, besides the well-remembered flak. Still, the B-17s pressed on, losing 19 percent of one day's force in the process.[21] Overall, the Ploesti B-24 raiders lost 30 percent of their aircraft during the day to AAA, fighters, fuel starvation from battle damage, and mechanical failure.[22]

North Vietnamese missileers used the Soviet-built SA-2 "Guideline" missile for defense against the LINEBACKER II raids.

It was a different matter for the LINEBACKER II bombers, and here we return to that most essential point: the LINEBACKER success was a team effort. Heavy preemptive strikes against the awesome enemy defenses were made continuously by F-111s and various combinations of Air Force, Navy, and Marine Corps fighter-bombers prior to the arrival of the B-52s. Special purpose F-4s sowed protective chaff while EB-66s and Navy and Marine EA-3s and EA-6s emitted electronic countermeasures (ECM) jamming signals to help hide the penetrating

CHAPTER 1 | PRELUDE

force.[23] F-105s, F-4s and Navy A-7s flew interspersed with the waves of bombers to deal on an immediate basis with ground defenses. Protective F-4s flew escort for the electronics aircraft and B-52s, while others flew combat air patrols to counter the fighter threat. The skies, already dominated by American airpower, were literally alive with friendly aircraft. It made for good odds, and the results prove that point. Only about two percent of the total force was lost, none to enemy aircraft, representing a dramatic improvement over previous efforts in similar conflict, and providing testimony to airmanship and systems capabilities.

The MIG-21 "Fish bed" was the primary North Vietnamese fighter aircraft. These aircraft sporadically engaged the bomber force during LINEBACKER II. B-52 gunners were credited with downing two of the defenders.

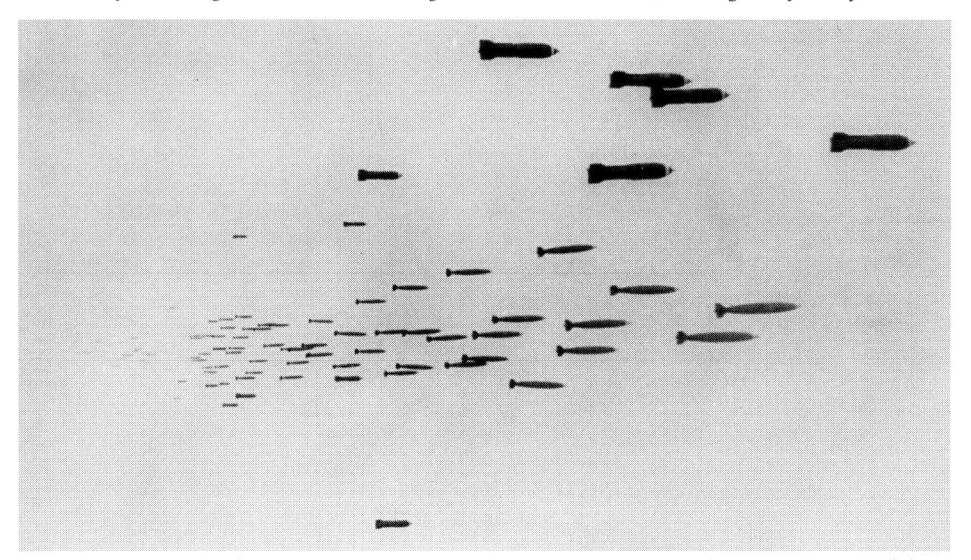

Bombs Away! Bomb bay view of 108 500and 750-pound bombs being dropped on a target obscured by an undercast. Precision electronic equipment enabled BUFF crews to attack targets in all types of weather and at any time of day or night.

9

NOTES

1. *USAF AIROPS*, p. IV-216.
2. Senator Barry M. Goldwater, "Airpower in Southeast Asia," Congressional Record, Volume 119, Part 5, 93rd Congress, 1st Session, 26 February 1973, p. 5346.
3. *Supplemental History on LINEBACKER II (18-29 December), 43rd Strategic Wing and Strategic Wing Provisional, 72nd (Volume I),* Air Division Provisional, 57th, Eighth Air Force, Andersen AFB, Guam, Mariana Islands, 30 July 1973, p. iv. TOP SECRET
4. John W. Finney, "B-52 Vindicates Its Role, Air Force Aides Assert," *The New York Times,* 24 December 1972, p. 3.
5. Henry A. Kissinger, "Vietnam Peace Negotiations," News Conference on 26 October 1972, *Weekly Compilation of Presidential Documents,* Volume 8, Number 44, 30 October 1972, p. 1566.
6. Message (TS), JCS to CINCPAC and CINCSAC, JCS 3348, for Gayler and Meyer from Moorer, 15/0147Z Dec 72 (72-B-7576). TOP SECRET. Subsequently declassified on 16 December 1977.
7. *History of Eighth Air Force, 1 July 1972-30 June 1973,* Volume I, Narrative-Part I, Andersen AFB, Guam, M.I., 23 August 1974, p. 147. Hereafter cited as *8AF History, V I.* SECRET
8. *USAF AIROPS*, p. IV-209.
9. R. Frank Futrell et al., *Aces and Aerial Victories: The United States Air Force in Southeast Asia 1965-1973,* The A. F. Simpson Historical Research Center, Maxwell AFB, AL, and Office of Air Force History, HQ USAF, 1976, Washington, GPO, 1977, p. 111.
10. *History of 43rd Strategic Wing, 1 July 1972-31 December 1972, BULLET SHOT Part II, With Emphasis on LINEBACKER II,* Volume I, Andersen AFB, Guam, M.I., 24 May 1973, p. 57. Hereafter cited as *43SW History.* SECRET
11. "N. Viet Earthquake: U.S. Bombs," *Pacific Daily News,* 30 December 1972, p. 4. See also *Omaha World-Herald,* 29 December 1972.
12. "Air Blitz On," *Pacific Daily News,* 20 December 1972, p. 1. Found also in many newspapers and periodicals from 19 December 1972 through February 1973, such as *The New York Times, the Omaha World-Herald, Newsweek Magazine,* etc. See also *USAF AIROPS,* p. IV-234.
13. *43SW History,* p. 58.
14. *8AF History,* V. II, Unnumbered charts in front of Chapter V.
15. Norman M. Whalen, "Ploesti: Group Navigator's Eye View," *Aerospace Historian,* Volume 23, Number 1, Spring/March 1976, Department of History, Kansas State University, Manhattan, KS, 1976, p.6. Used with permission.
16. *8AF History,* V. II, pp. 339-341.
17. *Ibid.,* p. 430.
18. Hanson W. Baldwin, "The Strategy of the Old Bombers," *The New York Times,* 19 January 1973. The theme is consistently projected both in contemporary newspaper articles nationwide and in such periodicals as *Aviation Week & Space Technology, Air Force Magazine,* and *Time Magazine.*
19. *B-52 Combat Damage Analysis,* Prepared by the Caywood-Schiller Division of A.T. Kearney, Inc., for the Joint Technical Coordinating Group for Munitions Effectiveness, Published as 61 JTCG/ME-75-1, October 1974, p. A-46. Hereafter cited as *Damage Analysis.* SECRET
20. *43SW History,* p. 100.
21. Thomas M. Coffey, *Decision Over Schweinfurt, The U.S. 8th Air Force Battle for Daylight Bombing,* New York, NY: David McKay Company, Inc., 1977, p. 332.
22. James Dugan and Carroll Stewart, *Ploesti: The Great Ground-Air Battle of 1 August 1943,* New York, NY: Random House, 1962, p. 222.
23. *USAF AIROPS,* p. IV-289

CHAPTER 2 | THE STAGE IS SET
BUILDUP OF THE FORCE

In early 1972, all B-52 sorties in support of Southeast Asia were flown by the 307th Strategic Wing at U-Tapao, known simply to most people as "UT." This was part of a long term involvement of B-52s in the Southeast Asian conflict conducted under the nickname "ARC LIGHT." Crews from all Continental United States (CONUS) B-52 bomb wings were assigned temporary duty (TDY) on a rotational basis to fly these missions. At the same time, the Andersen mission was to maintain B-52Ds and crews of the 60th Bombardment Squadron on nuclear alert.[1] The host unit, the 43d Strategic Wing, also was required to support an additional conventional warfare contingency plan for B-52s, if called upon to do so.[2] That contingency plan was soon to be exercised far beyond the parameters envisioned by its original drafters.

SAC's participation in LINEBACKER II didn't just "happen," and can only be fully appreciated by knowing something of what preceded it. Logistically, it could hardly have come at a more challenging time. The very factors which made the massive effort possible were the same ones which were applying seen and unseen stresses across the whole range of command activities.[3] At the local level, name any function, and in doing so you named a stress point. It ranged all the way from putting combat aircraft into the air to having enough beds to handle all the people. Andersen AFB was structured to function under normal duty hour conditions, and now base support organizations had to work 24 hours a day, seven days a week just to provide the minimum essential services.

No person or function escaped the impact, and it would not be far-fetched to say that this impact included the whole island of Guam.[4] Even the local housing market was affected, as some of the participants in the LINEBACKER II effort had brought their families to share the TDY experience—which by then had become more nearly the norm of their existence rather than the exception to it.

This broad impact had its roots in earlier events of 1972 which had continued unabated right up to the crucial days in December. Early in February, a squadron operations officer and crew at Carswell Air Force Base, Texas, were preparing to take the runway on what was supposed to be a routine night training sortie. After completing his checklist just short of the runway, the pilot had received takeoff clearance and started to taxi into position when he was ordered back to the parking ramp. When the crew reported back to squadron operations, they were directed to go home, pack their bags, and be back at the squadron in four hours. They were also told to be prepared to launch their B-52D on a long-range mission for an extended TDY. Thus was the beginning of operation BULLET SHOT, a systematic buildup of B-52 and support forces to counteract the increased infiltration pressure which North Vietnam was putting on the South.[5]

The 7th Bombardment Wing B-52D and others from Carswell were soon joined by those from the 306th Bomb Wing at McCoy Air Force Base, Florida, and the 96th Bomb Wing at Dyess Air Force Base, Texas. Their destination was the Rock. These men and aircraft were soon followed by their respective wing staffs and support personnel. This was to be but the beginning. Every B-52D unit in the CONUS would shortly follow suit, providing the expanded combat force of the 60th Bombardment Squadron and the 63rd Bombardment Squadron (Provisional).

The normal sortie length from U-Tapao was three and one-half hours. There was no refueling, and the crew duty day was approximately eight hours. The missions from Guam, on the other hand, required prestrike refueling and lasted approximately 12 hours. It was a more complex mission than those flown out of Thailand, and the crew duty day ran from 17 to 18 hours. This long mission and duty day meant that additional crew resources and tanker support had to be generated to support the Andersen sorties. Therefore, stateside tankers were sent to Kadena Air Base, Okinawa as the B-52Ds deployed to Andersen. Tanker operations also eventually expanded at several Thailand bases and in the Philippines, all of which would prove vital to the success of what was to come.

Normally equipped to accommodate around 3,000 personnel, and already supporting 4,000 due to the ongoing Southeast Asian effort, Andersen's ranks swelled month by month, until July when more than 12,000 people were sharing the real estate and facilities.[6] Many of these additional people supported their accompanying B-52G crews, who formed the Provisional 64th, 65th, 329th, and 486th Bombardment Squadrons of the activated 72nd Strategic Wing (Provisional). They brought with them nearly 100 "G" model B-52s from other units in the States, to add to the expanded force of approximately 50 "D" models. [Authors' note: the partial list of categories for classifying aircraft is: the "mission" (as in "B" for bomber), the "design" number (as in "B-52"), and the "series" (as in "B-52D," "B-52G," and so forth.) Therefore, the bombers would technically be referred to as the "D series" and "G series." However, it has become traditional within SAC to refer to them in daily life as "models." This text holds with the tradition.] A proportionate influx of equipment was incorporated to support the weapons systems and personnel needs. Those already present on the Rock adapted to this constantly and rapidly expanding challenge, allowing the war effort to go forward and the lives of the people to remain more or less tolerable.

CHAPTER 2 | THE STAGE IS SET

There were only two known SAC aircraft to sport shark's teeth during the Southeast Asian War. This KC-135 at Korat Air Base, Thailand, assigned to the 4104th Air Refueling Squadron, was an instant hit with Col Stanley M. Umstead, Jr. and the men of the 388th Tactical Fighter Wing.

LINEBACKER II | A VIEW FROM THE ROCK

Canvas Court #1. This is one of three tent cities built to accommodate the increased base population at Andersen. For the 12 men living in each tent, dust, noise, and tropical rains were a constant problem.

Because Andersen was not equipped to handle these 12,000 people who were assigned during the height of BULLET SHOT, improvisation became the order of the day. Overcrowding of normal living quarters had already occurred. Tent cities, recalling the days before and during World War II, sprang up . . . and stayed up. These remained for months as the visible expression of a modification of an entire base's operation. Just to house a portion of the additional support element required three of the tent cities, dubbed the "Canvas Courts," where each tent held 12 men.

To erect the tents was no simple matter. After digging six inches deep or less on much of the base, a shovel hits hard Pacific coral. To drive holes for tent pegs required jack hammers. Digging trenches for water and sewer lines was a major construction task. Once completed, the tents were little protection from the blanket of coral dust which settled over everything, and they offered no relief from the heat or the whine and roar of constant jet engine operations.

"Tin City," an Andersen landmark from an earlier era, became the overcrowded temporary home for thousands of additional maintenance and support personnel. The buildings, built of steel and corrugated sheeting, were intended as temporary spartan quarters for short duration TDYs. These H-shaped buildings, with a central latrine facility, were designed to hold 80 people while allowing the minimum floor space required by regulations for living quarters. Into these buildings were crammed 200 people. There was no air conditioning, and

CHAPTER 2 | THE STAGE IS SET

the only ventilation was from inadequate fans at each end of the bays. The temperature in these structures regularly exceeded 110 degrees in the noonday tropic sun. These were the sleeping quarters for those night shift workers who were loading bombs and maintaining the aircraft the crews flew in combat.

Even the tents and metal buildings, overcrowded as they were, did not provide enough space. Spare barracks space at the Agana Naval Air Station and Guam's Naval Base were pressed into service, as were the barracks at an abandoned US Army NIKE site. Still, the quarters were not sufficient. Spare rooms in all available hotels were rented, crowding four people into what would have normally been single accommodations. Crew quarters designed for two men held six.

The bus transportation just to haul these people to and from work amounted to more than seven million passenger miles a month. Although these off-base facilities were more liveable than the tents or Tin City, they were not preferred by the troops. Since everyone was on a minimum of six twelve-hour shifts per week, sleeping time became a very precious commodity. The two-hour round trip required to the off-base living quarters at the naval facilities or hotels meant 12 hours less rest per week.

Less openly seen than the tents and blue buses, but just as important to the people and the mission, were activities involving such diverse areas as the dining hall operations, the personnel center, legal office, recreation services, supply, etc. Even such amenities as the officers and noncommissioned officers open messes faced the task of providing services to huge numbers of people on an around-the-clock basis.[7]

"Tin City." These unairconditioned steel buildings were designed to house 80 men per building for very short periods of time. Into these buildings were crowded 200 men. Noonday temperatures inside these buildings exceeded 110 degrees.

Even the Base Gymnasium had to be used as living quarters during the buildup of personnel supporting BULLET SHOT.

Despite the crowding of six crewmembers into a living space designed for two men, there was not even enough room to house the entire crew force on base. Many had to be quartered off-base at whatever accommodations could be found. The lengthy round trip on the bus and problems of arranging for adequate meals extended their crew duty days to over 20 hours in some cases. Therefore, their insertion into the various training and flight schedules had to be carefully coordinated with their billeting arrangements, so that adequate crew rest was provided prior to flying the next combat mission.

The proper beddown of the crew force was crucial, for several reasons. Each organization had to have continuous and immediate control and access over its combat resource. A crew whose whereabouts could not be quickly determined might just as well not be on the island, due to the extensive coordination required to insert any given crew or crewmember into the flow of the operation. Considerations such as crew rest, individual crewmember substitution, and equitable distribution of exposure to combat were major parts of this coordination. Keeping track of as many as 216 crews, each with six individual crew specialists, often spread all over the island, was a challenging management problem. Because the crew force was a completely TDY contingent, there was the ever-recurring problem of monitoring rotation schedules. The CONUS-bound portion of this BULLET SHOT schedule did not apply during the LINEBACKER II experience, but there remained a continuing influx of those reporting back to TDY, plus the phased swapout of crews between Guam and Thailand. Those in the Crew Control Center, whose responsibility it was to manage this combat crew force, had their hands full.

AIRCREW TRAINING

Since the TDY crew force represented a cross section of experience in different type B-52s, the well-established training program was expanded to cover both models of the B-52 and to handle the increased buildup in crew numbers. If the crews who were to man B-52Ds came from B-52G or H outfits, they would first go to Castle Air Force Base, California, where they would receive a B-52D "difference" qualification check. This included both ground and flight training. When they finished this difference training at Castle, they would immediately be deployed to Andersen via Hawaii on KC-135 aircraft. Each tanker was usually packed with people and cargo. When they arrived at the Rock, the six-man crew would be jammed into any available crew quarters, on or off base. They were given 24 hours to complete in-processing and readjust their sleeping habits, attempting to reverse night and day in their systems. Then they attended a three-day "SCAT" school. SCAT was an acronym for Southeast Asia Contingency Air Training. In these three days the crewmembers would receive the latest changes to air tactics, survival procedures, updates on ordnance configurations, electronics countermeasures (ECM) changes, and the latest intelligence information on the areas over which they would fly.

After SCAT was completed, the crew would be scheduled for three indoctrination, or "over-the-shoulder," missions. On these missions a combat-seasoned pilot and radar navigator would accompany the crew. These experienced observers served a dual function; they offered advice to inexperienced crews if unusual conditions occurred, and they provided their squadron commander with an evaluation of a crew's potential for instructor or lead crew.

For pilots who had not regularly flown the B-52D, refueling was the biggest transition problem. The B-52D, compared to the B-52G or H, was underpowered and, because of cockpit seat positioning limitations, had poorer refueling visibility. Added to this was an increase in drag created by 24 external bombs which the D model carried on the underwing stub pylons. Therefore, a pilot who was able to get his full offload of fuel despite the loss of an outboard engine and an outboard spoiler group in the D model was considered to be a "real pro."

For the two-man navigation team, there was a different set of problems. The lower deck area in all models of the aircraft was basically the same, but equipment was differently placed and there were variations in the way certain components worked. Since special phases of the bombing mission involved intense concentration on the radar scope imagery, radar navigators (RNs) developed habit patterns based on the touch system. These patterns, deeply ingrained, broke down with different equipment characteristics and locations, and required much self-discipline and determination to overcome. The navigator (NAV) was less affected by equipment differences during the navigation phases. However, when he became a second set of eyes on the bomb run, he too had to insure that certain spontaneous reactions were in fact correct ones.

The elemental problem of equipment location posed similar problems for the defensive team as well. The most dramatic difference for the G model gunner transitioning to the D was being physically relocated from a position alongside the electronic warfare officer (EW) in the forward crew compartment to one in the tail end of the aircraft. Once transplanted back to his own special world, however, the gunner found the equipment very similar in operation. His defensive partner, the EW, like the RN and NAV downstairs, had to learn new switch locations and some equipment differences. He, too, needed to be able to operate systems by feel at those times when his display scope demanded full attention.

Major Robert A. Hewston, radar navigator veteran of three LINEBACKER II missions, cross-checks mission data at his downstairs position, nicknamed the "Black Hole of Calcutta." His cramped work area and maze of equipment are typical of B-52 crew positions.

CHAPTER 2 | THE STAGE IS SET

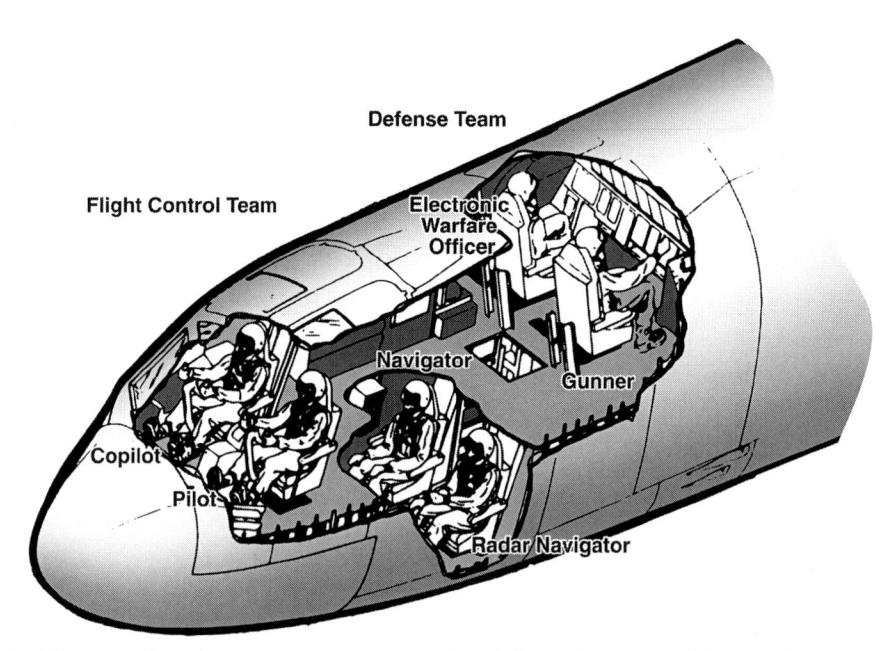

B-52G crew positions. The six-man crew is positioned in the forward 15 percent of the aircraft. All are equipped with ejection seats. The total area available for crew movement may be visualized by this sketch.

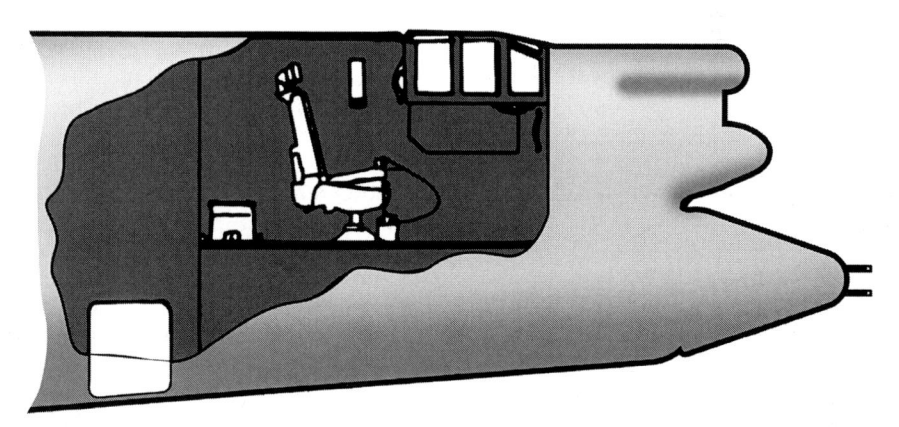

The B-52D tail gunner's compartment. Unlike the G model, five men occupy the forward crew compartment, with the sixth in the extreme tail end. This position has a gun turrent which may be jettisoned, followed by manual bailout.

If a crew showed normal progress, they received three over-the-shoulder flights—two during daylight and one at night. They were then cleared for solo. As they gained experience, crews were progressively qualified for "cell" and "wave" leadership. A three-ship cell was the standard B-52 strike mission formation, and a wave was a flexible combination of cells committed to a single strike objective. For those crews who demonstrated outstanding talents in both airmanship and teaching ability, there was the instructor crew designation. The best of the instructor crews were selected as tactical evaluation crews.

These tactical evaluation crews had differing and far more responsible duties than standardization evaluation crews did in CONUS units. Although they did fly with other crews on combat missions and observe crew performance, their primary responsibility was to evaluate the effectiveness of current combat tactics. After the observations and evaluations by the "Tac Eval" crews, new tactics were developed, if required. Once these new tactics were formulated, it became the task of those same crews to apply them in actual combat situations, and to insure that less qualified crews were capable of executing them. The tactical evaluation crews worked closely with combat operations staff officers to continually update tactics. This partnership would prove to be one of the key elements in. the success of LINEBACKER II.

THE "BICYCLE WORKS"

The success of BULLET SHOT and LINEBACKER II could not have been achieved without the tireless performance of the men of the 303d Consolidated Aircraft Maintenance Wing (CAMW). The "Bicycle Works" was a nickname used by the commanders for the CAMW. At the time, it was the largest maintenance organization in SAC, with more than 5,000 personnel, responsible at the height of LINEBACKER II for the maintenance of 155 B-52s. Approximately 4,300 of these men were on six month TDY tours. After a tour, they would get 30 days at their home base and then be sent back for another six months. When SAC war operations finally terminated in the autumn of 1973, many of these men were on their third consecutive six-month TDY to the Rock.

The sheer size of this complex maintenance operation made management a challenging task. Additionally, the time compression of maintenance actions called for the best efforts of maintenance supervisors. To meet the demanding schedule, planes had to be refueled and preflighted in a maximum of four hours—half the normal "stateside" time. Phase inspections were completed in eight hours—five times faster than normal, and many times were accomplished in the open during tropical rainstorms. There were only four nose dock hangars on Guam capable of handling B-52s. Two of these were on loan to a Boeing team performing mid-cycle depot maintenance. The two remaining were reserved for the most complex maintenance. Phase inspections, corrosion control, landing gear retraction tests, and fuel leak repair had to be done outside, day or night, in many cases on an unlighted ramp. A normal bomb wing engine shop would be hard pressed to overhaul five jet engines a month.[8] At Andersen, the requirement was 120 jet engine overhauls each month. The feats performed by maintenance technicians to insure on-time launches of aircraft became legendary. On one occasion, a tire change which would normally require two and one-half hours was accomplished in 15 minutes on the taxiway for a bomb-laden aircraft with engines running. On others, electronics specialists stayed aboard the aircraft to repair equipment in flight. These were voluntary gestures made without benefit of knowledge of the specific mission or possible hazards. It also meant an extension of at least 12 hours to what had already been a long, hard work day.

The tall vertical stabilizer of a B-52D dwarfs a maintenance technician in a "cherry picker."

As BULLET SHOT forces built up, the maintenance contract went from 21 to 66 B-52 sorties per day, plus required ground spares—seven days a week.[9] The reader should understand that 66 sorties would be the approximate number of training sorties that a normal CONUS bomb wing would fly in one month. To fly the 66 sorties, plus meet the demands of transient aircraft and tanker support sorties, required approximately two million gallons of JP-4 jet fuel per day.

Mutual respect between the flight crews and maintenance men reached new highs on Guam. The bomb wing commanders and maintenance wing commander developed programs allowing aircrews to attend maintenance roll calls to explain the purpose and results of the previous days' sorties. Beach parties were arranged so that flight crews and maintenance teams got to know one another better and understand each other's problems. Many crews took it upon themselves to invite the younger maintenance men to their individual crew gatherings. As a supplement to this program of interchange, those assigned to the bomb wing who expressed doubts about the dedication and effort of the maintenance personnel were given a week's TDY to the "Bicycle Works." This week produced amazing reversals in attitude.

"BAG DRAGS"

Each B-52 sortie and its success or failure was briefed at the highest levels of the Department of Defense. The Air Force and SAC had made a commitment to the ground commanders in Southeast Asia to provide a specific number of daily sorties. Commanders in the field were instructed to develop procedures which would insure that the B-52s provided all the sorties called for each day. This was accepted as a contract between SAC and the Southeast Asia commanders.

To minimize the possibility of a last-minute abort and loss of a sortie, a procedure commonly called "bag drag" was used. For each three-ship cell that launched, at least one ground spare would be preflighted up to the point of engine start. If something happened at the last minute to a primary aircraft, the crew would transfer to the spare and make good its scheduled takeoff time. For a B-52 crew to switch from one aircraft to another with all combat gear was no easy task. The amount of material a bomber crew had to take on a mission might seem absurd to other flyers. A fighter pilot might carry a map, mission data card, survival vest, sidearms, small packets of rations, and a checklist. For the heavy bomber mission, in addition to survival vest, sidearms, and large boxes of rations, a crew would have several oversize briefcases crammed full of classified mission materials, several full sets of aircraft technical manuals, bombing computation tables, celestial navigation data, other professional gear, and a full complement of cold weather flying clothes. It seemed somewhat ludicrous to see a crew haul cold weather gear out to an aircraft in the middle of the tropics.

CHAPTER 2 | THE STAGE IS SET

But loss of cabin heat was not an acceptable reason for an abort, and it happened from time to time. Flying six miles up for hours on end without cabin heat to offset an outside temperature which could go down to minus 56-degrees Celsius was impossible without this gear. It took two pickup trucks to handle a bag drag—one for the crew and one for its equipment. In some cases, a backup crew would preflight one spare aircraft and then go to another and preflight it. There were occasions where the backup crew would preflight, start engines, and taxi a third spare to an area just short of the runway, if required, to expedite a last minute bag drag and launch in as little as ten minutes.[10]

The bag drag was always a race against time in the heat or rain or both. Chop the power, set the brakes, throw the switches, grab everything and evacuate. Then followed a hurried ride to another aircraft, with its engines already running. Throw everything on board, strap in, and go—trusting that the preflight crew had completed all of the pretakeoff mission checks. Of all the acts involved in B-52 operations, none more fully showed forth the mutual faith in personal professionalism and job performance among the maintenance and operations people than did the "bag drag." The record for Andersen was set one hectic night when a crew made five successive bag drags and was able to join up inflight with its cell to complete the mission.

CHARLIE TOWER[11]

One of the most impressive activities associated with the SAC mission, whether at Andersen or U-Tapao, was the unique "Charlie Tower." It has had its rough equivalents elsewhere, such as the mobile control towers so familiar to pilot trainees. The Supervisor of Flying, a normal function at nearly every installation, might also be used as a basis for comparison, only to establish a common point of reference. The Charlie Tower was without comparison, by any standards.

It got the name "Charlie Tower" logically enough. The Wing Deputy Commander for Operations, third in the chain of command for operations, was responsible for the functioning of the tower. His radio call sign in most SAC wings is the phonetic "Charlie." Since the tower was the frequent location of the DCO, the association of the two became routine during radio calls, and the name stuck.

Over the years of SAC operations in Southeast Asia, Charlie Tower developed into more of a personality than a place. Its original concept must go down as one of the most superb techniques of on-the-spot controlling of masses of bomber aircraft ever developed, but there was more to it than that.

The people in the tower were what made it click, but there was a certain charisma about Charlie Tower which is difficult to put on paper. The "Charlies," who ran the tower for the DCO, were drawn from the most proficient of the crews. Said one seasoned SAC flyer:

They knew what it was all about. Even if the whole rest of the world was screwed up, they were the guys who had it together. When things were tense, puckered up, sometimes in a half-panic, the calm, methodical, unemotional, father-like voice coming from Charlie Tower was a stabilizing influence of great value.[12]

A base or unit command post has always been envisioned as the nerve center of any operation; and so it was on Guam and U-Tapao. But, once the scene had shifted out to the flight line for the launch and recovery phases, it was Charlie who choreographed one of the most impressive shows on earth. The ground movement of dozens of B-52s has been affectionately called a "parade of elephants" and the "elephant walk." Nobody could make them parade like Charlie. And since neither a B-52 nor KC-135 can back up of its own accord, the correct sequence of this parade is something requiring a discerning eye and forethought, if real embarrassment, or worse yet, a mission-crippling impasse due to a blocked taxiway is to be avoided.

To make all of this ground activity mesh, and to provide an immediate input on maintenance matters, other people were incorporated into the tower operations. In addition to the two Charlies, who were D and G model instructor pilots, there were "Uncle Ned" and "Uncle Tom." These were the Charlies' maintenance counterparts, not only in job specialty, but in capability as well.

Uncle Ned was the general overseer of maintenance activity in preparation for, and during, cell launches. He coordinated which tail numbers would fly, which aircraft were available for bag drags, and which maintenance problems were capable of being fixed by his "red ball" specialists. He could also draw on his own expertise or that of a systems specialist to assist a crew inflight with any problem they might encounter.

Uncle Tom directed the taxi and aircraft towing operations on the airdrome. During that period, the base was so packed with aircraft that it took an expert to insure their orderly movement from spot to spot. Much movement was necessary, due to limited facilities. One such limitation which impacted heavily on maintenance operations was the availability of refueling pits. Total time involved to fuel a BUFF was on the order of two hours, and there were ten times more B-52s than there were pits available to refuel them. Additionally, there were only certain parking spots which provided proper protection to other aircraft and personnel during 100 percent-thrust engine runs.[13] As a result, the taxitow operation was complicated, made even more so in that it had to be interwoven with the launch and recovery of a nearly steady stream of bombers, plus many transient fighter and cargo aircraft.[14]

The orderly movement of the fleet was so complicated that it was all Uncle Tom could do just to figure out how the jigsaw puzzle was going to work. To get it done, he further coordinated with a taxi controller known as "Cousin Fred." Fred had a pool of 12 taxi crews who went out and made Uncle Tom's wizardry come true.

CHAPTER 2 | THE STAGE IS SET

So there they were, one close-knit family: Cousin Fred working for Uncle Tom, who worked with Uncle Ned and the Charlies. This was all supervised by the final man in the tower, the ARC LIGHT Deputy Commander for Operations, who monitored the whole scenario. A visitor to the beehive known as Charlie Tower was always impressed, and properly so.

Out on the flight line, or within his radio control once airborne, Charlie was THE MAN. He enjoyed, deservedly so, unsurpassed confidence from the crew force. He was a sort of security blanket for the crews, especially the young pilots and those members of the G and H force who were flying the D model for the first times. If you wanted the straight word—"Call Charlie."

It took five miles of ramp space to park 155 B-52s, only about one-half of which are shown here. The Rock supported the largest concentration of B-52s in the history of Strategic Air Command. (Photo courtesy Capt Stephen D. Cross)

A typical situation would have a crew calling in with an aircraft problem which hadn't surfaced until after engine start. Since engine start was a function of launch and mission timing, the race against the clock was an automatic feature of any such problem. A quick assessment by Charlie, and one of several things happened. Uncle Ned might immediately dispatch a team of maintenance specialists.[15] Or Charlie and Ned would refer to their own comprehensive directory and point out an immediate fix which was within the crew's capability. Often, with that voice of wisdom and maturity which Charlies seemed born with, the original call for help would be answered with such words as, "No sweat, you're a 'goer.'" "Just isolate thus and so from the system," or "Just keep an eye on thus and so."

On the other hand, Charlie's knowledge of what was an acceptable aircraft and what wasn't enabled the tower team to make instant decisions on a "no-go" situation and coordinate a bag drag to another aircraft with no lost motion. Consider that it was routine for this type of activity to be going on while other aircraft were being launched, recovered, or towed. It took men of ability, poise, and self-discipline to man Charlie Tower. Such men filled the positions.

The respect and camaraderie that grew between the crew force and Charlie created some humorous situations, which make good war stories in the telling, but have meaning only to those directly involved. One of the more common occurrences, though, was the feigned formality of some crews returning from an exhausting all-night mission, who said, "Good Morning, Charles. Where shall I park?"

ARC LIGHT CENTER

To handle the deciphering of the fragmentary orders, or "frags," construction of the mission folders, radar target predictions, ECM information center and air intelligence, a special combat bomber operations center was established at Andersen. Under one roof were all the support elements necessary to plan the mission and brief the crews. Into this complex flowed the latest intelligence data on air and ground order of battle, last minute changes required to missions, and crew debriefing reports. If there was a heart to the air combat operation, Bomber Operations, or the ARC LIGHT Center, was surely it. For those personnel whose duty section was the ARC LIGHT Center, during LINEBACKER II it would serve both as work and sleep area.

Close companions to the ARC LIGHT Center planners were specialists from the Eighth Air Force staff, located just "down the road." Major Cregg Crosby remembers what LINEBACKER II and the preceding months were like for them and the Center occupants:

When did it begin? 18 December is the day that comes to mind—but, no that's not right, that's not even close if you were working in Eighth Air Force Directorate of Operations Plans, (8AF/DOX), Bombing and Navigation Division. LINEBACKER II really started in early August for us and thank God it did or it never would have been a successful military operation! Let me explain. Early August of 1972 we in 8AF/DOX began to hear the rumors of important efforts up North—rumor is really a misnomer. Actually, we received several classified messages of inquiry regarding our potential capability to wage this type offensive. Could we be ready? How many planes could be supported? How much refueling would be required? How about defenses and offset aiming points (OAPs)? The rumors in themselves were nothing new; we had been through these exercises more than once, but these messages had a sense of persistency to them. Then on 10 or 12 August we received a rather lengthy

CHAPTER 2 | THE STAGE IS SET

list of potential targets and work began in earnest. That list of potential targets proved to be amazingly close to actual LINEBACKER II targets. I would estimate that 90-95 percent of LINEBACKER II targets were on that original list. We did have subsequent refinements of target coordinates as new or better intelligence became available, but while the changes were numerous in quantity, they were minor in value. With this list of target coordinates we began many long, long days of offset aiming point/axis of attack selection.

This was a sidelight, so to speak, of the normal 105 sorties-per-day level attained as a result of BULLET SHOT (66 Andy and 39 U-T sorties). The rest of August, the three officers assigned to 8AF / DOX Bomb Nav worked days on

ARC LIGHT and late afternoons and evenings in the photo interpreters' vault, in an effort to select our recommendations for offset aiming points and axis of attack. With the aid of the photo interpreters we then built radar montages, and in a series of briefings over the next few weeks, presented these to the 8AF Commander, Lt General Gerald W. Johnson.

In the overall selection process, I guess we were lucky in some respects and not so lucky in others. Unlucky in that there was a great deal of SR-71 vertical photography to review, and this was a long, arduous task. The film was categorized by route and we received each and every roll. Much of that film was obscured by cloud cover, but generally speaking, when the ground could be seen the photo quality was excellent. We used this film and the stereo glasses extensively. We then used corroborating chart data to assist the selection of many OAPs. We were lucky, of course, around the prime areas of Hanoi and Haiphong. Prominent returns existed which would make good potential OAPs. Thus, as a result, prior to Day One almost all LINEBACKER II aiming points had been selected and briefed to the 8AF Commander at least once, and also forwarded to Hq SAC for approval. In fact by 15 December, Gen Johnson possessed excellent knowledge regarding our offset aiming point capability for the entire campaign and thus, our potential to accomplish accurate bombing for the first day's activity and for most of the targets throughout the 11day war.

As I recollect, it was probably fortunate that most OAPs were selected in advance, as things happened very rapidly after 15 December. For Day One the planners had adequate time to build the basic parts of the frags due to the previous stand-down, and except for delays imposed during the interpretation of certain last-minute directives, the frag might have been dispatched on time. This was the last time adequate planning time existed until approximately Day Five.[16]

As LINEBACKER II loomed on the horizon, the Rock was an armed camp of immense activity. A mass of humanity and its war machinery were firmly entrenched as part of the Andersen scene. It prompted inevitably the joke that the whole island was going to tip into the ocean because of all the weight at the north end.

HIGH THREAT PRESS-ONS

Zones of aerial operations are subject to a myriad of identifiers, based on the nature of operations. What is a hazardous environment to one type of aircraft in a specialized role may not even be remotely hazardous to another aircraft in the same air space. For simplicity of concept, B-52 combat operations took place in basically two zones—identified as low threat and high threat.

The dividing line between these two zones was primarily a function of the presence of SAMs of the SA-2 "Guideline" type. Added to this primary factor was the potential presence of Soviet built MIG aircraft and, to a negligible extent, AAA.[17] Of the latter two, only the MIG threat was sufficient to alter or modify B-52 mission planning and execution in the days prior to LINEBACKER II, and there were a few instances where such alterations did occur.

There was wide latitude in interpreting the low threat zone. It had gradations ranging all the way from no threat whatsoever to those borderline cases where B-52s were sporadically engaged by SAMs.

These unpredictable engagements had been taking place for a number of years, without success, and served mainly to keep the crews watchful and alert. As North Vietnam (NVN) solidified its holdings below the DMZ through 1972 and beefed up its bases of operation above the DMZ and in Laos, the frequency and accuracy of these sporadic firings increased.

Since the SA-2 system was readily transportable, missile firings could and did come from a wide variety of locations, sometimes on a one-time-only basis. Other firings would appear to have a pattern to them, which would occasionally persist, but might cease as abruptly as they had begun—not necessarily from the site having been destroyed.[18] This gave rise to the identification of suspected or probable operational areas for the SAMs, which was obviously the critical end of the low-threat spectrum.

It was only when missile firings had developed into a pattern or reached a level of regularity where they could not only be predicted but expected that the zone changed from low threat to high threat.[19] SAM activity now came from what was known as a confirmed operating area (COA), which was sometimes narrowed down even further to a confirmed operating location (COL). The latter remained a difficult determination to make in many instances, despite sophisticated intelligence efforts, because the NVN sensibly persisted in taking advantage of the mobile capability of the SA-2.[20]

As an aid to gathering and validating flight information, many of the aircraft were equipped with tape recorders, operated by the Electronic Warfare Officer (EW). These were selectively run during important phases of flight to record all communications, both from inside and outside the airplane.

The tape also recorded the audio signals which the EW was monitoring as he continually searched for friendly and hostile radar signals. During post-mission staff analysis, these

CHAPTER 2 | THE STAGE IS SET

tapes aided in confirming enemy threats by location and intensity, and helped to detect any changes in the enemy's air order of battle or in his tactics. These were correlated with the data which the planners were receiving from specialized electronics surveillance sources stationed in the forward areas to provide more comprehensive briefings. This information became extremely useful when developing new tactics against the more static defenses encountered during LINEBACKER II.

Throughout the summer and fall months of 1972, B-52 incursions into the borderline areas below the DMZ and high threat areas above it had been steadily increasing, accompanied by a commensurate increase in SAM firings. Even though effective interdiction of enemy supply routes and staging areas in this narrowest part of Vietnam was vital to long-range military and political goals at the time, the rules still provided for breaking off or diverting a mission if the anticipated or encountered threat became severe enough. In such a case, an alternate target, already preselected, would be struck. The NVN appeared to identify this tactic, and used periodic SAM or rocket firings effectively to force diversions away from their key positions.[21]

Captain C. T. Duggan, an EW from Westover Air Force Base, Massachusetts examines SAM damage to a right external tip tank, incurred during the November missions in the panhandle of North Vietnam.

During the second week of November, however, the concept of operations changed.[22] Diversions from primary targets were taking their toll in overall effectiveness while, at the same time, the enemy flow of supplies through the coastal and mountain pass routes above

29

the DMZ had reached alarming levels. Failure to stem this flow was no longer acceptable, and a heretofore rarely used tactic became the order of the day. For the crews from U-Tapao, Thailand, "Press-On" became as common a term as the "no sweat" cliché which it forcefully replaced.[23]

Reduced to its simplest definition, a "press-on" mission was one in which the aircraft continued to the target despite SAM or MIG activities in particular and aircraft systems degradation in general.[24]

Had the enemy fully realized the employment of the press-on concept from U-Tapao early on, the impact on the B-52 forces would undoubtedly have been greater. As it was, the incidence of near misses and limited battle damage increased, but this could be logically expected as a byproduct of the press-ons and not of concentrated defensive firepower.

Close-up of damage to the right external tip tank of Copper 3, on 5 November 1972. The SAM was so close that the gunner heard the motor of the missile prior to detonation.

As the crews at U-Tapao became acclimated to the daily ritual of flying press-on missions, the entire force was imbued with a heightened sense of professionalism. Closer attention was paid to premission and specialized crew position briefings. Aircraft and equipment were inspected and checked out in minute detail. Enroute cell procedures and tactics were thoroughly reviewed, and executed in flight with a precision overshadowing that of previous

CHAPTER 2 | THE STAGE IS SET

months. "Press-ons" were serious business, which business is the forte of the command. By the time of LINEBACKER II, the crews thus exposed to the press-ons were scattered among both the Andersen and U-Tapao forces, providing a nucleus of special experience and combat leadership.

Mated to the press-ons was another tactic, known as a "compression." By definition, a compression consisted of an average of three cells of three aircraft each striking the same target area with up to ten minutes' separation between cells. The compression tactic need not necessarily be tied to a press-on and had, in fact, been selectively employed on many occasions for saturation bombing effects.[25] In the cases at hand, though, it served the desired effect of saturating enemy defenses, increasing mutual ECM support, and simplifying the efforts of the supporting tactical and countermeasures aircraft.[26]

From the start, the guidance for the LINEBACKER II raids was that they would fly compression missions employing the press-on tactic.[27] This had deep implications for flights from the Rock, as the safety considerations for the long overwater haul were vital and less flexible than if the mission were confined to the land mass. The guidance to the crews up to that point, as related to safety, was that there was no combat mission important enough for the crew to accept any mechanical malfunction or weather condition that would risk the loss of an aircraft or the flight crew. There were specific malfunctions and weather conditions that made an aborted mission mandatory. There were other less serious conditions which were left up to the judgement of the aircraft commander (AC). Each AC knew what malfunctions he and his crew were capable of handling. A condition that might be safe for an instructor pilot, because of his flying ability and experience, might be unsafe for a new aircraft commander.

For LINEBACKER II, the rules were changed. The usual safety rules were suspended. Under the more stringent press-on rules of engagement, the aircraft would be flown if it were capable of delivering bombs to the target and recovering at U-Tapao. The loss of two engines enroute or complete loss of the bombing computers, radar system, defensive gunnery, or ECM capability were not legitimate grounds for abort. Consequently, there were only two air aborts from the Rock during the eleven days. One B-52, shortly after takeoff on December 26th, had to shut down all four engines on one side. With this condition, just flying the aircraft was a challenge, and delivering the bombs on target was out of the question. The pilot, Captain Jerre Goodman, patiently nursed the aircraft while remaining clear of the launch stream, and then elected to try for a landing back at Andersen. He would have been justified in bailing out the crew and heading the aircraft out over the water. Up to that time, the few situations involving landing a B-52 with all engines out on one side had been ones where a test pilot or highly qualified instructor pilot was aboard. Those had been accomplished under better flying conditions during daylight. Now, Capt Goodman had to attempt his landing at night. Despite this handicap, he did an outstanding job, made a safe landing, and thereby saved a much needed airframe.[28] The only other abort was caused by a complete failure of the refueling system on December 29th. With this malfunction, a B-52D did not have enough fuel to fly the mission and make it to an alternate base.

SPARE AIRCRAFT

The press-on rules did permit spare aircraft to be substituted on the ground up until scheduled takeoff time. Once the aircraft started its takeoff roll and attained a speed at which it could not safely stop on the runway remaining, press-on rules applied.

Two systems of "ground sparing" were used. At first, the "roll forward" principle applied. For example, if the number three aircraft in a three-ship cell had to drop out, the number one aircraft from the following cell rolled forward and took his place. Number two of that following cell then rolled forward and took number one's place as the cell lead. Three would become number two and a ground spare would become number three. Although this procedure expedited the fill-in process for an aborting aircraft, it negated some of the intracell planning done at the predeparture briefing. After the second day, a new concept was established. Spares were designated as backup for specific cells. For example, if number three aborted, a ground spare whose crew had attended that particular cell's briefing would substitute for the aborted aircraft. This substitution system worked better and gave the crews improved intracell coordination.[29]

All of the decisions to abort (terminate flight) after takeoff or to provide substitute aircraft were made by the wing commander of the aircraft involved. Because of the dedication of the maintenance people and the "can do" attitude of the crews, ground aborts were rarer during the eleven days than during the previous months.

A TASTE OF THINGS TO COME

It was inevitable. Some would call it the law of averages; others would call it blind luck. Others, for sundry reasons, would say we asked for it. But to the planners who were watching the scope and pattern of the November press-ons, and especially to the crews who were flying the missions, each passing day made it more and more obvious. Already, two aircraft had been safely recovered at U-Tapao earlier that month with varying degrees of battle damage, and there had been increasing numbers of SAM near-misses. At some point, the string had to run out.[30]

Lying roughly 150 miles above the demilitarized zone (DMZ), the target zone around Vinh was receiving primary attention from the November press-ons. It was likewise being defended in proportion to its importance as the largest industrial and transshipment complex in the southern part of the country, and as a focal point for the movement of supplies and

CHAPTER 2 | THE STAGE IS SET

equipment along the coastal network. It had long been a confirmed operating area, and SAM firings at B-52s in and around Vinh had become more numerous and more accurate.[31]

On the night of November 22d, Capt N. J. Ostrozny and his crew from Dyess Air Force Base, Texas, were flying as Olive 2 against one of the more frequently struck targets of Vinh. Immediately after bomb release, in a heavy SAM environment, the aircraft sustained lethal damage from a proximity detonation which punctured the underbelly and started fires in both wings and the aft fuselage.[32]

The payload of a SAM at the instant of detonation. Note the blotches and streaks surrounding the explosion. These are thousands of extremely hot warhead fragments, any one of which can be lethal if it penetrates a vital spot in the complex systems of the aircraft.

The crew concentrated on coaxing the steadily failing bird back to friendly territory. Capt Ostrozny made the decision to stick with the burning aircraft as long as possible. In the minds of his crew, a burning but flyable airplane seemed better than a prison camp. As the aircraft lost altitude, the gunner, SSgt Ron Sellers, who had lost communication with the rest of the crew, observed the fire in the right outboard engine pod and wing becoming brighter. The pilots could also monitor this progressing problem as they flew the aircraft towards Thailand. Then these and other engines started flaming out due to fuel starvation from the wing fires.

Sergeant Sellers, who could hear on interphone but not transmit, was determined to stick it out as long as he knew there was someone else in the aircraft with him—despite all that he was observing around him. Heat from the fire in the aft fuselage was becoming all too apparent in his tail compartment. He noted that the wing fires had turned a bright blue and had burned away part of the top of one wing. Soon he could see and count the internal ribs of the wing over the number eight engine pod.

Then the last engine quit. The pilot asked Capt Bob Estes, the navigator, how many more miles it was to the Thai border. The answer: five miles. The crew was alerted for an imminent bail-out. Now, for all practical purposes, they were a gliding torch, trying to trade precious altitude for those last long miles to safety. The airplane could explode at any second, Since the engines had quit, all electrical power was lost except battery power to the bail-out lights and interphone. The pilot lost his flight instruments.

The gunner observed the tip tank area starting to bend and fold up over the outboard engine pod. As the navigator announced that they had just crossed the Mekong River, the right wing tip broke off and the aircraft started an uncontrolled turn. Reacting immediately to the pilot's bail-out signal, the entire crew successfully abandoned their aircraft near Nakhon Phanom (NKP).[33]

Upon their return to U-Tapao, the crew gave detailed personal accounts of their experiences to meetings of crewmembers. One of these was taped, permitting their story to be related to the entire bomber force. Their safe recovery provided a happy human ending to the story, but now it was in the mind of each crewmember that the first Stratofortress had been lost to enemy fire after over seven years of combat operations.[34]

CHAPTER 2 | THE STAGE IS SET

NOTES

1. *8AF History,* V I, p. 173.
2. *43SW History,* p. 4.
3. *Ibid.,* pp. 1-3.
4. *8AF History,* V II, pp. 527-9, 557.
5. *43SW History,* p. 8.
6. *Ibid.,* p. 13.
7. *Ibid.,* pp. 155-158.
8. Colonel Myrell Hilger, Conversation with Lt Col Allison, 14 June 1977.
9. *43SW History,* pp. 22-23.
10. *History of 303rd Consolidated Aircraft Maintenance Wing (Provisional), 1-31 December 1972,* Andersen AFB, Guam, M.I., 20 January 1973, p. 25. Hereafter cited as *303CAMW History.* SECRET
11. *History of 72nd Strategic Wing (Provisional), 1 November 1972-31 January 1973,* Volume I, Andersen AFB, Guam, M.I., 28 April 1973, pp. 15-26. SECRET
12. Comment of U-Tapao pilot in 1971, recalled by author Allison; reconstructed in notes at Blytheville AFB, AR, Fall 1977.
13. *8AF History,* V. II, p. 590.
14. *303 CAMW History,* p. 24.
15. *Ibid.,* pp. 24-25.
16. Major Cregg Crosby, narrative written for the authors, 18 November 1977.
17. *Chronology of SAC Participation in LINEBACKER II,* HQ SAC/[HO], Offutt AFB, NE, 12 August 1973, pp. 25-26. Hereafter cited as *Chronology.* TOP SECRET
18. *USAF AIROPS,* pp. IV-174 to IV-175
19. *Chronology,* p. 27.
20. *43SW History,* pp. 99-100.
21. *USAF AIROPS,* pp. IV-47, IV-176 and IV-177.
22. *Chronology,* p. 15.
23. *USAF AIROPS,* pp. IV-49 to IV-50.
24. *History of Strategic Air Command-FY 1973,* Volume II, Narrative, Prepared by HQ SAC/ HO, Offutt AFB, NE, 2 May 1974, p. 225. SECRET
25. *Bomber ARC LIGHT Crew Manual,* 8 AFM 55-2, HQ 8 AF, Andersen AFB, Guam, M.I., 1 November 1972, Excerpts. SECRET
26. *SAC Southeast Asia Progress Report #84,* HQ SAC/[ACM] Offutt AFB, NE, November 1972. SECRET
27. *43SW History,* pp. 61-62.
28. *Chronology,* pp. 225-226.
29. *Ibid.,* p. 122. See also *303CAMW History,* p. 25.
30. *USAF AIROPS,* p. IV-54.
31. *SAC Southeast Asia Progress Report #84, op. cit.*
32. *Damage Analysis,* p. A-31.
33. Crew Narrative of 23 November 1972 Combat Mission, Cassette tape on file at A.F. Simpson Historical Research Center, Maxwell AFB, AL. SECRET
34. *43SW History,* pp. 48-49.

CHAPTER 3 | ACT ONE
THE DIE IS CAST

The result of the first warning order on December 15th was apparently slight. It had to be so. With the intended actions shrouded in secrecy, only a handful of people were involved.[1] Lt General Gerald W. Johnson, Eighth Air Force (8AF) Commander, was initially notified late on the 15th. The wing commanders subsequently got the word from the 57th Air Division (57AD) Commander, Brigadier General Andrew B. Anderson, Jr. Their preliminary instructions were to develop plans to have the wings prepared to mount a three-day maximum effort mission against heavily defended targets in North Vietnam.[2] They were to take no other action pending further instructions from 8AF.

Experience at wing level for such an undertaking was sparse. The guidance represented a radical departure from what the crews had been involved in for months. Although they had flown missions over North Vietnam, these had been restricted to the DMZ and the Southern Panhandle. Earlier in the year, the 307th Strategic Wing at U-Tapao had conducted a series of raids against Bai Thuong, Than Hua, and Haiphong. Colonel James R. McCarthy had been assigned TDY at the time to the 307th as an Assistant Deputy Commander for Operations. In that position, he was deeply involved with the planning and execution of those missions. He had also been assigned the project of reviewing the current MIG and SAM defense tactics and recommending changes to improve them. After the series of raids were completed, he was on the committee which critiqued the missions and made recommendations to higher headquarters. His experience was to prove invaluable during LINEBACKER II. There was only one other officer then assigned to either wing staff who had been involved in that series of missions.[3]

The morning of the 16th, there was a meeting at 8AF Headquarters, involving the division commander, wing commanders, and the 8AF staff. During the meeting, details were provided of the targets, routes to be flown inbound and outbound, and an in-depth description of radar aiming points for bomb release. There was considerable discussion on enemy defenses, supporting forces, and target area tactics.

Originally, the attendees were told to prepare for a three-day maximum effort.[4] All personnel rotations to the States were to be halted, and all available aircraft were to be put in commission for the upcoming raids. The decision was made to designate a full colonel as the Airborne Commander (ABC) for each day's strikes. He would be held responsible for the successful completion of the mission. Col Thomas F. Rew, commander of the 72d Strategic Wing, was selected to lead the first wave on the first day. Colonel McCarthy was selected to brief all three waves on the first day, plus the first wave on the second day. He was also designated the ABC for the first wave on the second day. Each colonel on the operations staff and wing commanders, including the commander of the 303d CAMW, were to take turns as ABC.

The most immediate task was to brief the wing staffs and the flight crews on how the missions were to be conducted. The wing staffs were briefed on the afternoon of the 16th and the flight crews were given a more limited briefing the next day. Lt Col Hendsley R. Conner, commander of the 486th Bombardment Squadron, recalled the situation surrounding the briefings:

Everyone was still hopeful that a truce would be reached in time for us to "get home in time for Christmas." All but a few bombing missions had been cancelled for December 17th. A meeting for all commanders was scheduled for 1400 on the 16th in the Eighth Air Force Commander's conference room. What was in the air? Were they getting the airplanes ready for us to fly home? As I left the squadron on my way to the meeting, I saw several crewmembers talking together.

One of them said, "Colonel, are we going home? Let's hope you have good news for us when you come back."

As we gathered for the meeting, speculation was running about fifty-fifty that we would be going home. Others of us had a premonition and were saying nothing. The General came in and the meeting got underway. The briefing officer opened the curtain over the briefing board, and there it was—we were not going home. No yet, anyway. We were going North. Our targets were to be Hanoi and Haiphong, North Vietnam. At last the B-52 bomber force would be used in the role it had been designed for. The goal of this new operation was to attempt to destroy the war-making capability of the enemy.

The method of attack we were to use would be night, high altitude, radar bombing of all military targets in the area of the two major cities in North Vietnam. We would launch a raid each night beginning on the 18th of December and continue with a raid each night. Each raid would consist of three waves of varying strength, each hitting their targets at four- to five-hour intervals.

It would not be easy. We knew we would suffer losses. The Hanoi/Haiphong target complex was among the most heavily defended areas in the world. The combined number of surface-to-air missiles, fighter aircraft, and antiaircraft guns that surrounded the target area exceeded anything ever experienced.

As soon as the meeting was over, I went back to the squadron area to begin preparations for the missions. As I approached the area, the crewmember who had wished for good

CHAPTER 3 | ACT ONE

news when I left was still there. He said, "We're not going home, are we? We're going North instead. I can tell from the look on your face."[5]

The pressure was on right from the start, and it was felt from top to bottom. This was far more than an all-out combat operation, which would have been pressure enough. It was to be a test of previously untested capability. The Chairman, Joint Chiefs of Staff (CJCS), Admiral Thomas W. Moorer, expressed his and the President's personal concern for the success of the campaign. It was abundantly clear—U.S. forces were expected, indeed obligated, to produce the desired results.[6] Further, their efforts would be monitored on a scale and with a level of concern unparalleled throughout the long years of the war. The sense of urgency and commitment to success permeated down through command channels undistorted, but reinforced. Consequently, by the time the word was received on the Rock, there could be no mistaking the challenge: take the ball and run with it, and don't stop until the goal is achieved. With the picture of SAMs and MIGs in the background, one could almost hear Admiral David Farragut calling to his helmsman, "Damn the torpedoes—full speed ahead!"

After the wing staffs had been briefed, there was considerable pre-mission planning which had to be accomplished. SAC Headquarters had planned the route to and from the target areas above the 20th parallel once the aircraft crossed north of the 17th parallel. This covered only about two to three hours of the 14-or-more-hour mission flight time. The rest of the routing and timing had to be planned by the wings and 8AF staff.[7]

The initial concept of operations as directed by the JCS called for around-the-clock bombing of the Hanoi-Haiphong area. Tactical fighter and fighter-bomber forces from Seventh Air Force and comparable aircraft from the Seventh Fleet would strike during the day and the B-52s would strike at night. The first night's sorties would consist of 54 B52Gs and 33 Ds from Andersen, plus 42 Ds from U-Tapao. These 129 planes would fly in three different waves, with approximately four to five hours between waves. Strikes would first be made by the U-Tapao crews against Hoa Lac, Kep, and Phuc Yen Airfields, suppressing MIG-21 "Fishbed" operations, while the rest of the force attacked Hanoi area targets, coming in generally from the northwest. This cone of attack was selected to insure ready identification of radar aiming points and give minimum exposure time to the lethal SAM defenses deployed in the target area. After the aircraft had dropped their bombs on target they would turn and exit the target area on tracks heading back to the west and northwest.[8] The second and third days were to be repeats of the first day, except the total number of aircraft would be 93 for Day Two and 99 for Day Three. The axis of attack and withdrawal routings were to be essentially the same for all three days, but comparison of the mixture of targets spread over the three days shows that there was considerably more variety to ingress and egress routing than there was first believed to be. The one constant in all strikes during the three days would be the post-target turn (PTT).

LINEBACKER II | A VIEW FROM THE ROCK

18 DECEMBER 1972, WAVE 1

B-52 CELLS/TARGET TIMES

'D' GUAM		'G' GUAM		'D' U-TAPAO	
ROSE	2001	RUST	2007	SNOW	1945
LILAC	2003	BLACK	2009	BROWN	1947
WHITE	2005	BUFF	2011		
		CHARCOAL	2014	MAPLE	1949
		IVORY	2016	GOLD	1951
		EBONY	2018	GREEN	1953
				PURPLE	1955
				WALNUT	1957

LEGEND

- - - - - - - CHINESE BUFFER ZONE
　　　　　　　APPROXIMATE SAM COVERAGE
　　　　　　　TARGETS
　✈　　　　　BOMBER ROUTE IN
　　　　　　　BOMBER ROUTE OUT
COLOR　　　 CALL SIGN OF CELL

TARGETS

1 HOA LAC AIRFIELD	6
2 KEP AIRFIELD	9
3 PHUC YEN AIRFIELD	6
4 KINH NO COMPLEX	18
5 YEN VIEN RAILROAD	9
	48

39 SUPPORT AIRCRAFT

EB-66 ECM
F-4 CHAFF
F-4 CHAFF ESCORT
F-4, B-52 ESCORT
F-105 IRON HAND

CHAPTER 3 | ACT ONE

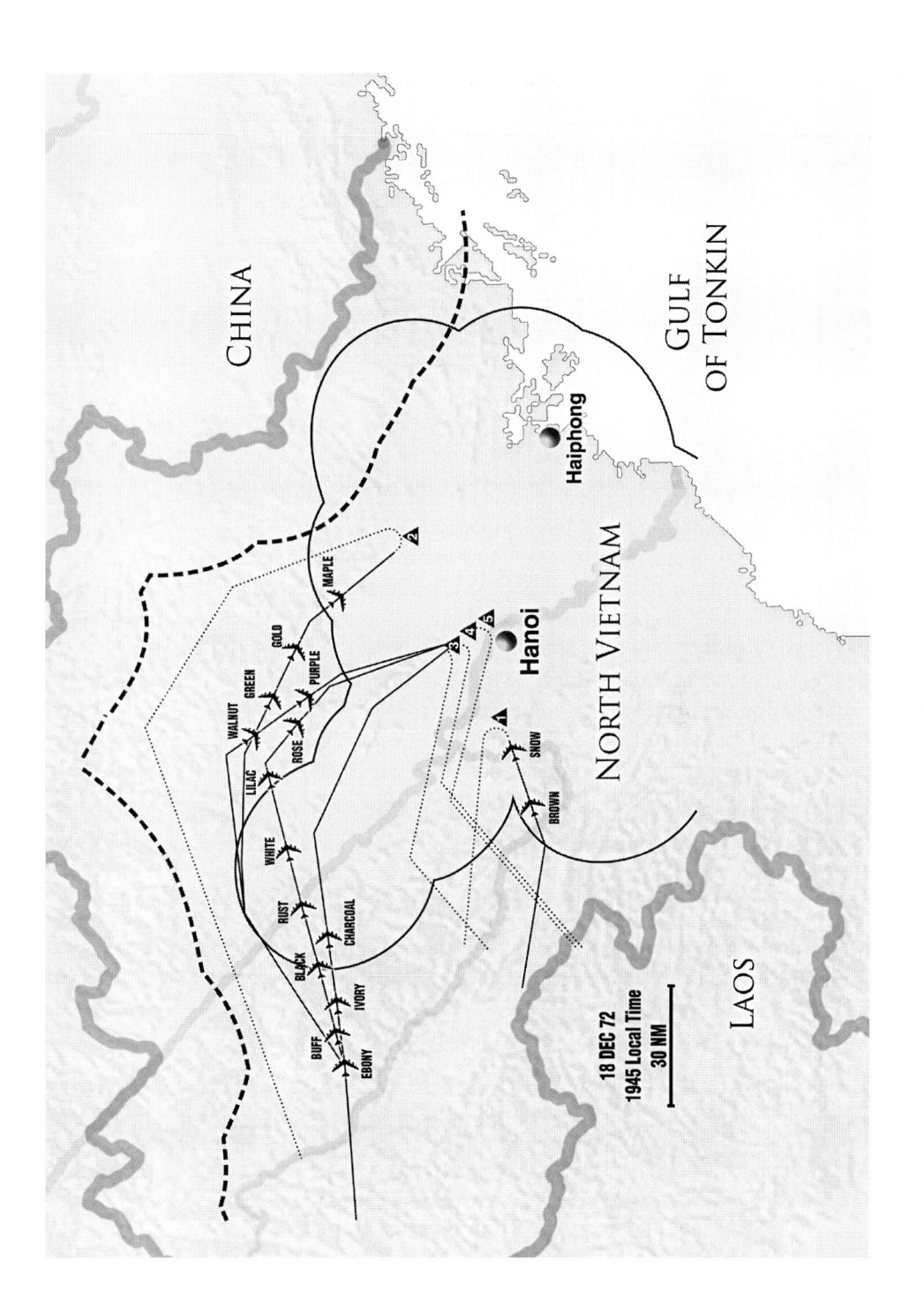

There was a key point of debate within the staff as to what would happen after the three-day maximum effort. If there was to be a day's standdown afterwards, another maximum effort could be launched on the fifth day. The unknown tasking for the fourth day presented a dilemma. The three days would tax the crews and the maintenance structure, but by going to 18-hour shifts they could handle it. If, on the other hand, the maximum effort were to continue, then work hours needed to be planned differently. After considerable discussion it was decided to plan for an extended period of days with maximum effort missions. This would later prove to be a wise decision.

Another priority order of business was selecting the crews who would fly cell and wave lead. To lead the first D model Stratofortresses in the first wave, the commander selected Major William F. Stocker's crew from McCoy Air Force Base, Florida. They had been involved in the Haiphong raid earlier in the year. Before he came to SAC to fly BUFFs, Bill had flown in F-4s over NVN. He had seen SAMs in quantity on that tour. In addition to those combat missions, he had approximately 300 in the B-52, and was thus well qualified to lead the raid.[9]

The experience of the other crews was analyzed in similar detail to insure that the most experienced and combat-proven were in the critical leadership positions. This was easier said than done. Prior to any knowledge of LINEBACKER II requirements, the planners had worked out a flying schedule that would allow some additional crews to go home for Christmas. They were able to do this because other crews had volunteered to fly more than their normal share of sorties during the holidays. The feeling among the flight crews who had to stay on the Rock was that if they couldn't get home themselves, they could at least fly additional sorties to allow more of their fellow crewmembers to spend the holidays with their families. It had made for admirable sentiment, but difficult scheduling.[10]

After the initial guidance was given to the staff, a meeting was held for aircraft commanders in the ARC LIGHT Center briefing room. Others were excluded simply because the briefing room was not nearly large enough to hold all the crewmembers concerned.

Among the aircraft commanders filing into the room were several who were scheduled to rotate back home the next day. Many of the others had hoped that the meeting was being called to tell them the war was over and that they, too, could spend Christmas with their loved ones. Expectations were high because the newspapers and TV had been full of rumors about a possible breakthrough on the peace talks in Paris.

These men had to be told that they and their crews were all restricted to the base until further notice, including those with families on the island. No outgoing or incoming calls would be allowed. Most of the missions for the rest of the day would be cancelled, but they should be prepared to fly a series of maximum effort missions on the 18th. They couldn't be told what the targets were at this time, but were informed that press-on rules would apply.

CHAPTER 3 | ACT ONE

Those who were already familiar with flying press-ons took that news in stride, but in the restrictions they recognized a clue to something big.[11]

After instructions had been presented to them, the first question asked was, "Sir, for those of us whom the Flight Surgeon has grounded, do you have any objections if we try to talk him into 'ungrounding' us?" Such was the morale of this outfit.

Some DNIF (duty not involving flight) crewmembers did talk the flight surgeon into letting them fly. Of those, three were later personally cited for heroism and valor in these missions.

Flying high altitude missions with a blocked ear or sinus can be extremely painful, and can lead to ruptured ear drums or serious infections. Sometimes the individual involved thinks the infection minor and doesn't find out how serious it really is until he flies the mission. Before LINEBACKER II was over, Colonel McCarthy was to learn the hard way the penalty for flying combat missions with double pneumonia.

There were a thousand and one details that had to be checked and rechecked on the 17th. The latest enemy air order of battle was examined in minute detail. Bomb fuse settings and time intervals between bombs had to be computed and verified. Aircraft availability, crew schedules, transportation, food service, and life support equipment had to be coordinated. The first planning data on the missions showed that the B-52Ds would be marginal on fuel reserves, even with the inflight refueling. The normal fuel reserve for a B-52 returning to Andersen was 30,000 pounds over the initial approach fix at an altitude of 20,000 feet. This would be enough fuel to allow the aircraft to be inflight refueled again in the event there was a landing gear problem or other malfunction, provided a KC-135 tanker could get off the ground within five minutes after notification.

Now the mission planning for these lengthier flights into the North showed that the aircraft would return with just 20,000 pounds of fuel. This may seem a lot of fuel but, as the jet bomber pilot knows, once you descend from high altitude with only that amount of fuel you had better "have it on the deck" shortly thereafter. Just to add spice to the situation, the only "deck" available within three hours flying time was an apparent dot in the middle of the ocean, the island of Guam.

Another important consideration was how to handle battle damaged aircraft. Because of the long mission duration there would be times when Andersen would be recovering one group of bombers while simultaneously launching another from a second runway. When this happened, both runways had to be open at all times during the combined activity.[12] Closing the runway to recover a badly damaged aircraft or one with hung ordnance was not feasible. The decision was made to recover aircraft of questionable status at Northwest Field. Northwest Field was an old World War II bomber base with an 8,000 foot runway. Narrow and short for B-52 operations, it was the only other field available, save Agana Naval Air Station, which also served as the civilian airport for Guam. Agana's use could have serious implications for commercial traffic; hence the Northwest Field alternative. Fortunately, during the entire eleven-day operation this emergency landing strip was not used.

LINEBACKER II | A VIEW FROM THE ROCK

As December 18th approached, the magnitude of the new operation began to have its impact on the support structure. The normal base support for BULLET SHOT operations was geared to the cyclic launch of three aircraft and recovery of three more every hour and one-half.[13] This meant, for example, that normally six crew buses, plus two for the ground spare crews, supported daily operations. Likewise, the inflight kitchen had to prepare only 18 flight lunches every hour and one-half to support the crews.

Now the requirement for the first night's launches would be 87 buses for the primary crews and equipment, plus six for the spare crews. There were only 50 buses available to support the entire base. Buses were also needed continually to haul the thousands of TDY people back and forth between their on- and off-base quarters, dining halls, and the many work areas scattered across the sprawling base. The inflight kitchen suddenly found itself with a requirement for over 500 flight lunches at one time, rather than the normal steady supply of 18 every 90 minutes.

Security Policemen maintained a constant vigil, guarding the largest assemblage of combat-ready B-52s and supporting KC-135s ever gathered on one base.

Other support functions such as bomb loading, refueling pit operations, dining halls, security police, civil engineering, and life support found their capabilities stretched to the limit. The logistics problems already created by BULLET SHOT escalated to crisis proportions. It was in these areas of combat support that the leadership, imagination, and ingenuity of the dedicated NCOs and officers paid off. Hundreds of individuals made on-

CHAPTER 3 | ACT ONE

the-spot decisions at all levels to get the job done. Never once did anyone say he couldn't support the surge.

To insure that support for this unprecedented effort was closely coordinated, a special support command post was set up under the leadership of Colonel John Vincent, who had taken command of the 43d Combat Support Group in October.

On the morning of the 18th, the staff reviewed the combat tactics and made sure the support package was developing according to schedule. Tactically, SAC Headquarters had directed that no maneuvering to avoid SAMs or fighters would be allowed by the bombers from the initial point (IP) on the bomb run to the target.[14] Such maneuvers, called the "TTR" for their effect against target tracking radars, were a part of ARC LIGHT operations. The order not to maneuver, which puzzled the crew force, was predicated on a number of factors, one of which was concern over mid-air collisions—either among the bombers or the bombers and support forces. Of more importance was the need for mutual ECM support, which required cell integrity. Most important of all, in this first stab at the enemy's vitals, was the political-military concern over the proximity of many strategic targets to population centers. This required the flight crews to be doubly sure they were on the planned course and in a trail formation, so that the train of bombs from each aircraft would impact along the desired path on the ground. The stabilization systems for the bombing computers aboard the aircraft required a certain amount of straight and level flight to properly solve the bombing problem; otherwise, the bombs might be scattered outside the target zone. As the missions progressed, and analyses of accuracies could be made, this amount of straight and level flight might be reduced, if circumstances dictated. However, accuracy and assured destruction were overriding considerations. Bombers on the first raids were required to stabilize flight for approximately four minutes prior to bomb release. It was a long four minutes. Unfortunately for some, it lasted less than that.

Keeping the formation together was an important consideration. The ECM equipment on the B-52 is designed to give the aircraft protection from SAMs, especially when the aircraft in the formation are mutually supporting one another. This protection is basically achieved by creating receiver interference on the enemy's radar and trying to confuse his missile guidance signals. The bomber tries to hide the aircraft radar "signature" in this interference, or receiver jamming, as it is called.[15]

Without jamming, an aircraft appears as a clear, small blip or spot on the radar scope. When jamming is introduced, this blip is covered with a wide band of interference. If the effort is successful, the ground controller doesn't know where to aim his missile. Success depends on the B-52s staying in formation. Three aircraft the size of a BUFF can put out a lot of jamming power. However, if one aircraft gets out of formation, one-third of the cell's mutual protection is lost, making it easier for the enemy to find both that aircraft and the others on his radar. Worse yet, the aircraft that becomes separated from his cell stands out like a sore thumb because his jammers highlight his position. In this case, the enemy also has the capability to go to a "track on jam" mode of operation. That results in the missile being commanded to follow the jamming signal to its source—the aircraft.[16] Mutual ECM support by maintaining intracell position was to become a key element in LINEBACKER II.[17] For an aircraft the size of a B-52, it is a challenging task to maintain the exacting formation positions that afford maximum ECM protection, but the experiences of those who did not maintain such positions were positive proof that it was a challenge worth meeting.

As a supplement to the jammers which hid the attacking force,[18] additional jammers were employed to disrupt communication between the SAM site and its launched missile.[19] The controlling ground radar emitted a command guidance signal known as an UPLINK, to which the inflight missile responded with a beacon signal known as a DOWNLINK. Desirably, both were jammed by the penetrating aircraft.[20]

UPLINKs became a way of life, and an EW's staccato announcement of "SAM UPLINK" over the interphone caused hearts to flutter, and botched up many a train of thought before the campaign ran its course.

Preloaded bombs on "clip-in" racks. Bombs can be loaded one at a time, or preloaded and inserted as a clip-in package, if special preload equipment is available. There was no preload equipment at Andersen; all bombs had to be loaded in the bomb bay racks one at a time. Shown is a full internal load of 84 500-pounders.

As the morning of December 18th wore on, the seemingly insurmountable problems were being progressively solved. The solution to the crew bus situation, for example, was to put a

CHAPTER 3 | ACT ONE

cell of three crews on each bus. These were the small GI buses, but by packing every nook and cranny with crew equipment, including the aisles, it was possible to get three crews and their bulky gear aboard. Additional required life support equipment was borrowed from the Naval Air Station. Volunteer off-duty cooks were pressed into service to handle the surge of flight lunches. By tripling the maintenance effort on the out-of-commission aircraft, sufficient ground spares were available at launch time.

Throughout the entire Andersen complex the story was the same. The importance of the mission, plus the motivation and leadership at all levels, produced the extra effort necessary to insure the successful launch of the required combat sorties. Significantly, that spirit of extra effort would never wane on the Rock.[21] Bomb loaders who had put in twelve exhausting hours of struggling with heavy bombs in the hot tropic sun refused to leave their duty stations. Instead, they stayed on and gave the second shift the needed manpower to complete the uploads.

In the ARC LIGHT Center, the mission planners and target intelligence people were busy preparing the crew materials that would be needed that night. These were hundreds of the most qualified specialists in the command. They would soon know the mental strain of trying to deliver a perfect product to the crew force, with the silent prayer that pressure and cramped working conditions had not allowed serious error in their computations. Many were also veteran crewmembers of exceptional knowledge and maturity, who continued to voluntarily fly combat missions to insure familiarity with the combat tactics. They also asked to fly as extra crewmembers on the LINEBACKER II missions. Those gestures were vetoed, the odds of survival being poor for those without ejection seats when a sweptwing jet sustains a major hit.

For these planners, last-minute changes to routes, enemy order of battle, and unusually strong upper level wind systems became major problems. No sooner would they complete and reproduce the mission materials than a late frag order change would come over the wire that would invalidate three or four hours of work. This would prove to be a continuing problem during LINEBACKER II.[22] As the briefing time for the first mission rapidly approached, it became apparent that, due to another last-minute change, the crew target study materials would not be ready in time for the scheduled crew briefing.

Thirty minutes prior to the first scheduled briefing time, additional changes to the frags were received that even made it necessary to delay the general briefing by one hour. However, the master battle plan and the coordination with in-theater forces made no provisions for delays like this. So, staffers bowed their necks and processed the data just that much faster. Word was passed to the flight line that the crews would be running behind time. It would be a hectic few hours. Nevertheless, the national command decision to strike had been made—everyone pressed on.

47

LINEBACKER II | A VIEW FROM THE ROCK

DAY ONE—HOW DO YOU LIKE THE SUSPENSE?

Reflecting on the momentous briefing for Wave 1 on the 18th, General McCarthy recalls:

As the crews filed into the briefing room there was the usual milling around and small talk between crewmembers, made the more so by the large numbers of people.

Since I was selected to give all three briefings on the first day, I tried to come up with some words or phrases that would convey the message of the importance of the targets to our national goals; yet I wanted to keep it simple and uncluttered. After a few minutes' debate with myself, I chose the simplest of opening statements. As the route was shown on the briefing screen I said, "Gentlemen, your target for tonight is Hanoi." It must have been effective, because for the rest of the briefing you could have heard a pin drop. Having sat through some 1,200 predeparture combat briefings myself, I can truthfully say that that group of combat crews was the most attentive I have ever seen.[23]

"Your target for tonight is Hanoi." These crewmembers show the tension, drama, and concentration which characterized the briefings for the LINEBACKER II missions.

48

After the general briefing, the crewmembers attended specialized briefings. The pilots' specialized portion included intracell coordination, communications and procedures to be followed if one of the aircraft in the cell was shot down or damaged. The RN and NAV closely examined radar predictions of the release offset aiming points for their specific target, and vertical photography of the target area. The instructions to the RNs were that if they were not 100 percent sure of their aiming point, "then don't drop; bring the bombs back."

The EWs were given the latest information on the position of the enemy SAMs, early warning radars, ground control intercept radars, possible airborne fighter interceptor radars, and heat seeking missiles plus other classified enemy emitters.

The gunners were concerned with the possible threat of enemy fighters. In the B-52D, the gunner's compartment is like a miniature cockpit in the rear of the aircraft, from which he has an easy field of view of about 270 degrees. He has both an optical and a radar sight for his four .50-caliber tail guns. The gunner on a B-52G, on the other hand, sits in the forward crew compartment next to the EW. He has a radar gun sight, but no optics. The G model was originally configured with a remote TV camera, but that feature was deactivated about the time the Gs were first deploying to Guam. Even if it had worked, it was no 'substitute for a set of eyeballs in the back of the aircraft, and the ability of the D gunners to monitor other aircraft and SAM firings and flight paths to the rear of the aircraft was a significant advantage for their crews.[24]

Close-up view of the "D" tail gun system. Guns were pointed up during maintenance. The gunner's canopy resembles fuselage-mounted pilots' canopies of older style aircraft. For bailout, the turret is jettisoned at the seam just aft of the canopy.

There was lively discussion on the MIG and SAM threats among the crews and staff. One group thought MIGs would be the greater threat and, therefore, tactics should be maximized against them. The other side argued that SAMs would be the greater hazard and, therefore, everything possible should be done to keep the formation together for mutual ECM support. One aid that could be used for this purpose was the upper rotating light. This red flashing light could be seen for four to five miles by aircraft-flying above the lighted craft. The weather forecast for Hanoi was supposed to be low overcast conditions, which would prevent enemy gunners below from visually tracking the aircraft.

The wing commanders directed the crews to use the rotating lights periodically as an aid to keeping the formation together. If any MIGs were reported, then all aircraft would turn off their lights.

Once the first briefing was completed, the next task was to launch the aircraft. At this point, the nerve center for the whole operation moved across the field to the Charlie Tower. For this launch, the limited room inside the tower was especially cramped. In addition to its normal complement of people, there were the Commanders of 8AF, 57AD, 43SW, and the 303 CAMW. The 72 SW Commander was the Airborne Mission Commander.[25]

The first aircraft started its roll at 1451 local time. Major Bill Stocker's black-bellied D model slowly lumbered onto the runway. As the nose of the aircraft lined up with the white center stripe, he advanced the eight throttles to takeoff power. Black smoke belched out the tailpipes as water augmentation was added to increase takeoff thrust. Slowly the BUFF started to accelerate. The engines not only had to overcome the inertia of a 450,000-pound aircraft, but also the 150-foot dip and rise from the middle to the end of Andersen's swaybacked runway. The airspeed seemed to climb ever so slowly, but soon had passed "decision speed"—the point at which momentum prevents the aircraft from being safely stopped on the remaining runway. From this point until the aircraft reached takeoff speed, as in all B-52 launches, there was little the crew could do but "sweat it out." A blown tire, loss of power from outboard engines, or any other major failure could spell disaster for the crew and aircraft. With all systems "in the green," (i.e., indicating normal), Major Stocker eased back on the control column and Rose 1 took off. Seconds later it was over the ocean. Twenty-six Ds and Gs followed in close succession and the Rock's contribution to Wave I was airborne.[26] LINEBACKER II had begun.

CHAPTER 3 | ACT ONE

Lt General Gerald W. Johnson, a fighter ace of World War II, was the Commander of 8th Air Force during LINEBACKER II. Here he intently watches launches from the vantage point of Charlie Tower.

LINEBACKER II | A VIEW FROM THE ROCK

Lt General Gerald W. Johnson, 8AF Commander (right), and Brig General Andrew B. Anderson, Jr., 57AD Commander, gathered in Charlie Tower along with the rest of the senior staff to monitor the complicated taxiing and launch of scores of airplanes.

Its landing gear starts to retract as a fully loaded B-52D launches off the north end of Andersen's runway enroute to Hanoi.

CHAPTER 3 | ACT ONE

As the Stratofortresses started down the runway at Andersen, the rest of Wave I at U-Tapao were undergoing last-minute preparations to begin launch. Supporting forces at other bases in Thailand and aircraft carriers in the South China Sea were launching their aircraft as part of the operation. Navy fighter bombers would attack coastal gun emplacement sites, plus known or suspected SAM batteries. Specially configured F-4s would lay down chaff, the metallic or fiberglass strips of reflective tape which degraded the enemy radar scopes. Before the campaign was concluded, they would dispense 125 tons of this deceptive material.[27] Along with the chaff would be EB-66s, EA-3s, and EA-6s, packed with ECM equipment which would create additional clutter on enemy radar screens and further hide the bombers.

Other F-4s, F-111s, and A-7s would be attacking airfields and SAM sites along the B-52 ingress and egress routes. F-105 Wild Weasel aircraft would be searching for enemy radar signals against which they would launch their SHRIKE radar homing missiles.[28] Other F-4s would join up with the bombers to provide the MIG CAP fighter protection. EC-121Ts would be providing special monitoring capability that would warn the strike force of close-in threats. Rescue C-130s and HH-53 "Super Jolly Green Giant" helicopters were on station, ready to provide immediate recovery help to the strike force if required.

At sea, a specially designated Navy vessel of Task Force 77 (TF 77)[29] off the coast of Haiphong was standing by to issue SAM warnings, vector fighters against any MIGs, and assist damaged aircraft with escorts or directions to reach the nearest friendly ships offshore. This vessel was known to the flight crews as "Red Crown," a friendly voice to count on for help when needed.[30]

The briefing for Wave II was conducted in the midst of the flurry of excitement surrounding the Wave I launches, but it went smoother than the first briefing. By now the shock effect was over. Additionally, the weather, intelligence, and target briefers had polished their presentations.

The first crisis came as the prestrike refueling report was received from Wave I. Major Stocker reported 196,000 pounds as his total fuel after receiving the briefed offload from the tanker. He should have reported 216,000 pounds. He was sitting with 20,000 pounds less than planned, and that was exactly the fuel reserve specified back over Guam. By the book, the mission should have been aborted at this point. Without additional fuel, and flying the mission as briefed, the force would arrive back home with dry tanks. If they were diverted into bases in Thailand after the strike, the aircraft could not make it back to Andersen to be regenerated in time to meet the second and third days' mission requirements.

To abort the mission would be playing it safe, and no one could be criticized for following the book. Yet there was the collective effort, with all its implications, to consider. If the mission was continued and something went wrong or the force wasn't able to meet the next days' requirements, then responsibility would shift directly to the source of that decision.

There wasn't much time to make the decision; within minutes the strike force would start to turn north. The commanders discussed the alternatives in a hastily called meeting. The only way to complete the mission and have the aircraft available for the next day's mission was to get them more fuel—from somewhere. Although there were KC-135 tankers available in Thailand, they were already committed to pre- and post-strike refueling of the F-4s, F-105s, and EB-66s that were supporting the bomber missions. The only possible tankers available were at Kadena Air Base, Okinawa. These, except for a few spares and strip alert aircraft, were also committed to the refueling of the follow-on B-52 waves over the South China Sea. To use them and then insure they were available for quick turnarounds to support the follow-on strike aircraft, the post-target refuelings would have to be close to Kadena. This would mean that a 14-hour mission would become an 18-hour one for some of the bomber crews. Another complication was that the weather in the vicinity of Kadena at refueling altitude was predicted to be marginal at best.

Generals Johnson and Anderson questioned whether the crews were experienced enough to make this type of major inflight modification to their route, then post-strike refuel in marginal night weather conditions to complete the mission. The big unknown, however, was what condition they and the airplanes would be in after the bomb run. How many would have battle damage, injuries, hung ordnance, or be exhausted? Weighing those questions, Col McCarthy expressed confidence that his crews could handle it. The generals decided to continue the mission. Word was flashed to the force to expect poststrike refueling over Kadena for those which were low on fuel.

A quick check of the flight plans revealed that the headwinds, which were much stronger than predicted, were responsible for the difference between the planned and actual fuel reserves. The B-52Gs, having a larger initial fuel load and more fuel-efficient engines than the D models, could overcome this adverse condition and would not require a second refueling.

To insure that this situation would be covered on future missions, the 376th Air Refueling Wing at Kadena was instructed by 8AF to rearrange aircraft scheduling to include sufficient tankers for post-strike refueling if required. This was a major change to the 376th's original tasking. It required additional quick turnaround sorties, more fuel offload per aircraft, and longer duty days for the tanker crews. The 376th, under the leadership of Colonel Dudley G. Kavanaugh, met all additional taskings every day. During the LINEBACKER II campaign, not one sortie was lost due to lack of refueling support. This achievement was realized despite persistent marginal weather conditions, last-minute changes to refueling areas, and numerous mechanical malfunctions.

Each briefing for the first raids became more polished. By the time the specialists briefed Wave III, the presentations were the professional "dog and pony" shows SAC crews had

CHAPTER 3 | ACT ONE

come to expect. There were also fewer changes to the frags, and the target material section was able to present some of the crews with their own mission folders prior to the briefing.

B-52D #678, stabilized in the pre-contact position, prepares to hook up with a KC-135A for inflight refueling. Number 678 was severely damaged on 18 December while flying against the Kinh No Complex as Lilac 3. (Photo courtesy Lt Col Glenn Smith)

Lieutenant Colonel George Allison, who had flown a 12-hour interdiction mission near Quang Tri, South Vietnam, the night of the 17th, viewed Day One from the perspective of a crewmember who could only watch what was unfolding:

> We all knew something was up—there was no way to hide it. The previous standdown, during which only a few of us flew, and then the abrupt gearing up for action could only mean one thing. A list of fliers for the day as long as your arm, cell color designators many had never even heard of before, a parking lot so full of crew buses that it looked like a staging area, hundreds of your contemporaries threading their way through the main door of the ARC LIGHT Center, plus a day of rumors, made it all too clear. The whole force was set in motion to do something big, and it didn't take a very shrewd person to figure out what. There were only two places deserving of such feverish activities Hanoi or Haiphong. The only thing lacking was the official stamp of authenticity—the formal word at the pretakeoff briefing.
>
> Those of us not assigned to the first day's missions never heard that formal word, but we didn't need to. It was almost a foregone conclusion, shortly given substance by the crewmembers departing the first briefing.
>
> It is difficult to describe a feeling which develops gradually around intuition, hunches, rumors, logic, and so forth, but subsequently draws its substance from fact. Was it happiness, relief, a sense of "we're finally going to do it; it's long overdue"? Or was it dismay,

confusion, a sense of "has it actually come to this?" I can't remember, and I'm satisfied that I can't because all of these feelings were intermingled in the fascination of the moment.

Glad that we were finally going to show them a thing or two? Yes. Somewhat relieved that I wasn't initially one of those "we"? Frankly, yes. Would there be another day like this? Perhaps not. Maybe once ought to get the point across.

The mood around the crew quarters was elusive. I wasn't the only one sharing a confusion of emotions. Many were going through the same experience. The rumor mills and speculative bull sessions ground on, and people looked at one another with peculiar expressions. It was all too much to take in, in one swallow.

Ever so slowly, the reality and significance started to sink in. As the first of the flight crews prepared for launch, clusters of people started to gather here and there. Many took advantage of the balconies on the multi-storied crew billets to watch the scene unfold. It was an occasion for conversation, poor jokes, more rumors, and sober reflection. Somebody out there wasn't going to come back, and we all knew it. The vantage point from the crew quarters lent itself to the drama of the occasion. The runway was plainly visible, and each launch could be watched up through a point at which the crew was positively committed to a takeoff. That point of commitment, essential to every takeoff, took on special meaning as we watched.

The most impressive sight, however, was the preliminary to the launch. For, from this same vantage point, the only portion of most of the taxiing force which was visible was the vertical stabilizers. They could be seen moving along the line of revetments, an assembly line of aircraft tails. The similarity of these to the moving targets in a shooting gallery stuck in my mind. I didn't like the analogy, but it was too vivid to dismiss. They would move forward, ever so slowly, but the line seemed to never stop. A constant procession of aircraft tails, almost too much to grasp. Where were they all coming from to get in a continuous line like that? When would the line end? We don't have that many B-52s on the whole island!—which can be the sensation when you've lost count.

Finally around midnight, they were all airborne and the ensuing silence was as thunderous as the hours of launches. It was time to grab a bite to eat, do a lot of thinking, and take a quick glance at the next day's list of flyers. Probably wouldn't be anything going after an effort like this, but checking was part of the required routine. That was a bad guess on my part, and the next day it would be my turn.[31]

After the launch of the third wave came the hard part on the Rock—sweating out crews and aircraft. A lot of questions go through your mind while you are waiting. How many will we lose? Will the weather hold up in the post-strike refueling area? How many aircraft will have hung bombs—always a hazard on landing? How many casualties will we have? In some respects it's easier for any commander to fly the mission than to remain behind. At least when you are flying the mission you know what is happening to your outfit and how they are doing.

CHAPTER 3 | ACT ONE

Crewmembers, dressed for the tropics, gathered on the balconies of the aircrew billets to watch their buddies launch on Day One.

LINEBACKER II | A VIEW FROM THE ROCK

One commander found out all too clearly what was happening to the outfit. Lieutenant Colonel Conner, flying a G model (Peach 2) tells his story:

Each squadron was given the responsibility for one wave of each raid. The staff and I worked almost around the clock getting things ready for the first raid. The schedule was prepared, crews were notified, transportation ordered, flying equipment prepared, meals ordered, and all the myriad of things that must be done to prepare 33 crews and airplanes for a combat mission.

Each wave would have a senior officer along as an Airborne Mission Commander. The wing commander would be the ABC on the lead wave and I was assigned as Deputy ABC to the second wave. I would not fly in a crew position, but would go along in the instructor pilot's seat as the seventh man on the aircraft. By not having any crew duties, I could concentrate on how the mission was progressing and be aware of any problems the wave might encounter.

The planning was complete, the briefings were finally over, and we arrived at the aircraft to preflight the bombs and equipment. Wave II was scheduled to begin taking off at 1900. Every 90 seconds after the first takeoff another fully loaded B-52 would roll down the runway. Anyone who has ever witnessed such an event can never forget it.

After we leveled off, I tried to get some sleep. I had gotten very little rest the night before because of the many problems that had come up during mission planning. I slept about three hours before the copilot woke me up for our inflight refueling. Since the mission was scheduled to last over 14 hours, refueling was necessary in order to complete the mission and land back at Guam.

When the refueling was over, I tuned in the radio to hear how the lead wave was doing. They should then be in the target area, and we should be able to hear how the enemy was reacting. The first report I heard was when Col Rew made his call-in after they exited the target area. They had had a tough experience. One airplane was known to be shot down by SAMs, two were presently not accounted for, and one had received heavy battle damage. He initially estimated that the North Vietnamese had fired over 200 SAMs at them. There were no reports of MIG fighter attacks. The anti-aircraft artillery was heavy, but well below their flight level. For us, the worst part was now they knew we were coming, and things probably would be even worse when we got there.

I saw the SAMs as we came in closer to the target area. They made white streaks of light as they climbed into the night sky. As they left the ground, they would move slowly, pick up speed as they climbed, and end their flight, finally, in a cascade of sparkles. There were so many of them it reminded me of a Fourth of July fireworks display. A beautiful sight to watch if I hadn't known how lethal they could be. I had flown over 200 missions in B-57s

CHAPTER 3 | ACT ONE

and I thought I knew what was in store for us, but I had never seen so many SAMs. I did not feel nearly as secure in the big, lumbering bomber as I had in my B-57 Canberra that could maneuver so much better.

Just before we started our bomb run, we checked our emergency gear to make sure everything was all right in case we were hit. We would be most vulnerable on the bomb run, since we would be within lethal range of the SAMs and would be flying straight and level. We had been briefed not to make any evasive maneuvers on the bomb run so that the radar navigator would be positive he was aiming at the right target. If he was not absolutely sure he had the right target, we were to withhold our bombs and then jettison them into the ocean on our way back to Guam. We did not want to hit anything but military targets. Precision bombing was the object of our mission. The crews were briefed this way and they followed their instructions.

About half way down the bomb run, the electronic warfare officer on our crew began to call over the interphone that SAMs had been fired at us. One, two, three, now four missiles had been fired. We flew straight and level.

"How far out from the target are we, Radar?"

"We're ten seconds out. Five. Four. Three. Two. One. BOMBS AWAY! Start your . . . turn, pilot."

We began a . . . right turn to exit the target area.

KABOOM! We were hit.

It felt like we had been in the center of a clap of thunder. The noise was deafening. Everything went really bright for an instant, then dark again. I could smell ozone from burnt powder, and had felt a slight jerk on my right shoulder.

I quickly checked the flight instruments and over the interphone said, "Pilot, we're still flying. Are you OK?"

"Yes, I'm fine, but the airplane is in bad shape. Let's check it over and see if we can keep it airborne. Everybody check in and let me know how they are."

"Navigator and Radar are OK. We don't have any equipment operating, but I'll give you a heading to Thailand any time you want it."

"EW is OK, but the Gunner has been hit. We have about two more minutes in lethal SAM range, so continue to make evasive maneuvers if you can."

"Gunner is OK. I have some shrapnel in my right arm, but nothing bad. The left side of the airplane is full of holes."

I called the lead aircraft to let them know we had been hit. He said he could tell we had been hit because our left wing was on fire and we were slowing down. I asked him to call some escort fighters for us.

The airplane continued to fly all right, so the pilot resumed making evasive maneuvers. We flew out of the range of the missiles, finally, and began to take stock of the airplane.

The SAM had exploded right off our left wing. The fuel tank on that wing was missing along with part of the wing tip. We had lost #1 and #2 engines. Fire was streaming out of the wreckage they had left. Fuel was coming out of holes all throughout the left wing. Most of our flight instruments were not working. We had lost cabin pressurization. We were at 30,000 feet altitude. Our oxygen supply must have been hit, because the quantity gauge was slowly decreasing. I took out two walkaround bottles for the pilot and copilot. If we ran out, they, at least, would have enough emergency oxygen to get us down to an altitude where we could breathe.

We turned to a heading that would take us to U-Tapao, Thailand. I called again for the fighter escort to take us toward friendly territory.

"We're here, buddy."

Two F-4s had joined us and would stay with us as long as they were needed. One stayed high, and the other stayed on our wing, as we descended to a lower altitude and to oxygen. They called to alert rescue service in case we had to abandon the aircraft. Our first concern was to get out of North Vietnam and Laos. We did not want to end up as POWs. We knew they did not take many prisoners in Laos.

Thailand looked beautiful when we finally crossed the border. Since Thailand was not subject to bombing attacks, they still had their lights on at night. We flew for about thirty minutes after we had descended to a lower altitude, and began to think we would be able to get the airplane on the ground safely. The fire in the left pod was still burning, but it didn't seem to be getting any worse. One F-4 left us. The other one said he would take one more close look at us before he, too, would have to leave. His fuel reserve was running low. He flew down and joined on our left wing.

"I'd better stay with you, friend. The fire is getting worse, and I don't think you'll make it."

I unfastened my lap belt and leaned over between the pilot and copilot to take another look at the fire. It had now spread to the fuel leaking out of the wing, and the whole left wing was burning. It was a wall of red flame starting just outside the cockpit and as high as I could see.

I said, "I think I'll head downstairs."

"Good idea," said the pilot.

CHAPTER 3 | ACT ONE

The six crewmembers in the B-52G have ejection seats that they fire to abandon the aircraft. Anyone else on board has to go down to the lower compartment and manually bail out of the hole the navigator or radar navigator leaves when their seat is ejected. I quickly climbed down the ladder and started to plug in my interphone cord to see what our situation was.

The red ABANDON light came on.

BAM! The navigator fired his ejection seat and was gone.

The Radar turned toward me and pointed to the hole the navigator had left and motioned for me to jump. I climbed over some debris and stood on the edge of the hole. I looked at the ground far below. Did I want to jump? The airplane began to shudder and shake, and I heard other explosions as the other crewmembers ejected. I heard another louder blast. The wing was exploding. Yes, I wanted to jump! I rolled through the opening, and as soon as I thought I was free of the airplane, I pulled the ripcord on my parachute.

I felt a sharp jerk and looked to see the parachute canopy open above me. The opening shock felt good even though it had hurt more than I had expected. Everything was quiet and eerie. There was a full moon, the weather was clear, and I could see things very well. I looked for other parachutes. One, two, three, that's all I saw. Then I saw the airplane. It was flying in a descending turn to the left and the whole fuselage was now burning and parts of the left wing had left the airplane. It was exploding as it hit the ground.

I saw I was getting close to the ground, so I got ready to land. I was floating backwards, but I could see I was going to land in a little village. I raised my legs to keep from going into a hootch. I certainly didn't want to land in someone's bedroom. I got my feet down, hit the ground, and rolled over on my backside. I got up on one knee and began to feel around to see if I was all right. Everything seemed to be fine. There was a little blood on my right shoulder from where a piece of shrapnel had hit, but otherwise, just bruises. It felt good to be alive.

About twenty or twenty-five Thai villagers came out of their homes and stood watching me. They were very quiet and friendly and brought water for me to drink. None of them spoke English, so we spent our time waiting for rescue, trying to communicate with sign language. They kept pointing to the sky and showing what must have been an airplane crashing and burning. I tried to describe a helicopter to let them know one would be coming soon to pick me up. I hoped.

In about twenty minutes a Marine helicopter did come, and I was picked up. We had bailed out near the Marine base at Nam Phong. All six of the crew had already been rescued, and none had serious injuries. We were flown to U-Tapao, and then on back to Guam the next day. Our particular ordeal in the bombing raids was over. The crew had performed well; I was proud of them. The reason I had decided to fly with them on the mission was because I thought they were one of my most professional crews. They were from the 2nd Bomb Wing, Barksdale AFB, Louisiana. The aircraft commander was Ma jor Cliff Ashley. His copilot was Captain Gary Vickers. The radar navigator was Major Archie Myers, navigator 1st Lieutenant Forrest Stegelin, electronic warfare officer Captain Jim Tramel, and gunner Master Sergeant Ken Connor. The outstanding way they handled our emergency showed how competent and courageous they were.

The crew I had flown with, along with survivors from other aircraft, were flown back to the States for rest and leave. They were short of squadron commanders on Guam, so I had to stay and help prepare other raids that were continuing each night.[32]

The weather at Andersen for recovery was perfect. Each aircraft would report his battle damage, fuel reserve status, and maintenance malfunctions. As the reports came in to Charlie Tower, all were amazed. The D models were coming back with very little battle damage and very few maintenance discrepancies. Fuel reserves were close, as had been predicted. As an extra precaution, several strip alert tankers had been launched to orbit near the penetration point in case an aircraft developed a serious last-minute malfunction.

Listening to the intelligence specialists debrief the crews at many tables which had been set up in the gymnasium, it was painfully apparent that all was not "peaches and cream." There had been problems on the mission. Some aircraft, contrary to instructions, broke formation to dodge SAMs. Other crews reported problems in knowing the exact location of other aircraft in the cell and their MIG CAP. Some crews lost their radar equipment and had problems exiting the target area. Fighters and bombers had similar sounding call signs, or transmitted instructions without identifying themselves, which created confusion on the radios. Radio frequencies in the target area became unusable because of oversaturation of transmissions concerning SAMs, MIGs, and AAA.[33] Some aircraft had malfunctions in their release systems. Worst of all, but expected, were the lost aircraft reports.

The very first Stratofortress to go down in LINEBACKER II, Charcoal I, provided one of those tragic and poignant stories which are a special part of war.[34] Lt Col Donald L. Rissi and his crew from Blytheville Air Force Base, Arkansas, had originally been scheduled to rotate back to the States on December 4th for their one-month break. Their replacement crew had become snowed in at Loring Air Force Base, Maine. When that crew finally did arrive, they were minus two primary crewmembers. This required extra over-the-shoulder training, which slipped Colonel Rissi's "go-home" date to December 18th. Scheduled to fly east, he flew west instead.

Don Rissi's navigator, Capt Bob Certain, related a pretakeoff sequence for Wave I which was both humorous and ominous. During engine start, Rissi candidly observed that if all of the aircraft were to line up and advance their power at the same time, they "could probably break this end of the island off." Shortly thereafter, while taxiing, an urgent call came from the

CHAPTER 3 | ACT ONE

control tower: "All B-52 aircraft, stop taxiing!—earth tremor in progress!" The old volcanic mountaintop was up to her usual tricks again, protesting mildly at all of the noise. It was of short duration, but still was not one of the signs one hopes for in such a "terse" situation.[35]

Colonel Don Rissi, who by all reasonable odds should have been at home two weeks earlier, lost his life from a SAM detonation only seconds before the bomb release point over the Yen Vien Railroad Yards.[36] His gunner, MSgt Walt Ferguson, died with him. The copilot, 1/Lt Bob Thomas, joined the ranks of those missing in action (MIA). The radar navigator, Maj Dick Johnson, gained international news "exposure" via a picture of him being marched in his underwear through the streets of Hanoi,[37] where he joined Bob Certain and the Electronic Warfare Officer (EW), Capt Dick Simpson as prisoners of war (POW).

Through the course of eleven nights of intense combat, this crew's fate would be shared in varying degrees by 14 other crews, two of them that very night. Total losses were the two Gs from Andersen and Capt Hal Wilson's U-Tapao D model (Rose 1). Two Ds from Andersen had received extensive battle damage and were forced to recover at U-T.[38]

Compensation for this sobering fact was the realization that 94 percent of the night's force had released on their targets,[39] and that the enemy, too, had paid a price in the air. SSgt Sam Turner, flying in Brown 3, made his own history by being the first gunner in a B-52 to down an enemy aircraft in combat.[40]

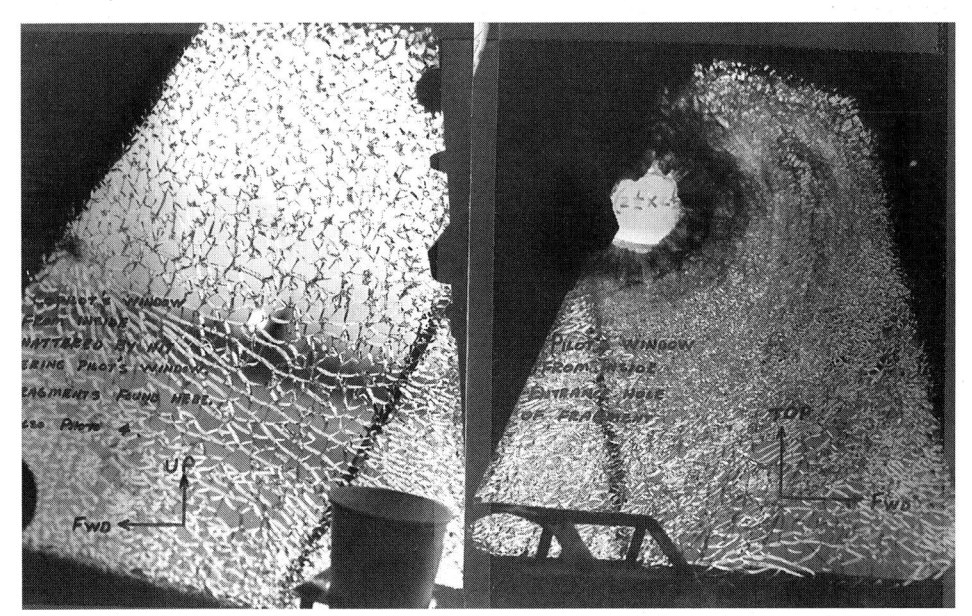

The pilots of Lilac 3 (#678) on 18 December 1972 were fortunate. A SAM fragment penetrated the pilot's side window and passed directly across the cockpit, shattering the copilot's side window before it spent itself.

DAY TWO—REPEAT PERFORMANCE

In reality, it was impossible to separate Day One from Day Two on the Rock; the same would hold true for Day Three. The "all night" bombardment concept and the long drive back and forth across the ocean effectively made it a continual effort. The last cells of Day One were landing while the first bombers of Day Two were starting engines.

Buildings, supplies, and rolling stock lie in ruins along the Kinh No Railroad Spur.

All that was humanly possible was being done to assimilate and consolidate the mass of data coming in from the crew debriefings of the first attacks so the findings could be passed on to the second day's flyers. Most efforts focused on the avoidable problems which surfaced from the first crews' experiences, and on a recapitulation of what to expect. However, even the more complex issues were immediately set upon by all available experts.[41]

Time compression made it impossible to clear changes in tactics through the higher headquarters. Except for different targets in several instances, and varying weights of effort, Day Two was to be Day One all over again. It was at this point that General Anderson gave both wing commanders permission to change certain air tactics as they thought best for their individual wings. This decision allowed adjustments to refine combat tactics in meeting the continuing changes occurring in the air battle.

CHAPTER 3 | ACT ONE

The changes in tactics and their execution during the eleven days of LINEBACKER II were significant achievements of the campaign. As the raids progressed, so too did the complexity of the tactics. No one on Day One or Day Two could visualize the dramatic changes in combat tactics which would unfold in the days immediately ahead. Still, pilots with as few as four missions as aircraft commanders would be executing these maneuvers flawlessly over downtown Hanoi.

Maj Tom Lehar from Dyess Air Force Base, Texas, was selected as the leader of Wave I on the second night. Tom had impressed the commander and staff with his "can do" attitude and professionalism as a pilot; his crew would be joined by the ABC as the seventh man.

As with the first night's missions, the aircraft in Wave I were not allowed to maneuver from IP to target. That would change before the night was over for the rest of the force.[42] The crews and staff were disturbed by the entry and exit routes for the second night's missions because of their similarity to the first night. Also, as in the first night, the aircraft were required to make turns to exit the route after bombs away. Recommendations were made to SAC Headquarters that permission be granted for the bombers to maneuver until just prior to the bomb release point. Suggestions were also made that the ingress and egress routes be changed so that a pattern was not set which would make it easy for the enemy to preposition his forces.

The Hanoi Fabrication Plant, with its surrounding support buildings and supplies, shows the effects of the intense bombardments directed in the Kinh No area.

After analyzing the debriefing comments from Waves I and II on the first day, Colonel McCarthy was convinced that mutual ECM protection was the key to reducing losses. For this reason, he gave his 43d Strategic Wing's crews a precaution that night that was to be unpopular—that he would have to consider court-martialing any aircraft commander who knowingly disrupted cell integrity to evade SAMs. It didn't please some crews, who believed that they could dodge a SAM. It was one of those unpopular decisions a commander must make for the good of his command. Its wisdom was to be tested in the crucible of battle.

A "D" model taxis into position for launch as a "G" model lands in the background from the previous day's bombings.

The target for Wave I was again the Kinh No Railroad and Storage Area on the outskirts of Hanoi. The strike force consisted of 12 B-52Ds and nine Gs from Andersen. The Ds in this wave carried forty-two 750-pound M117 bombs internally and twenty-four 500-pound MK 82 bombs on the underwing pylons. The Gs carried their smaller internal load of 27 M117s.[43] It was a persistent sore point with the G crews that their unmodified aircraft only carried that much striking power, especially when they had to penetrate the same defenses. Also, because the G was still a relative newcomer to the conventional mission, release system malfunctions plagued that model.[44] More than once a crew was unable to release even a single bomb.

During the pilots' specialized briefing, the ABC and aircraft commanders discussed some of the feedback of the pilots from the first day's missions who had been debriefed only hours before. Since MIGs hadn't been a problem and the weather was forecast to be low undercast, the upper rotating beacons would again be used as an aid to formation flying.

CHAPTER 3 | ACT ONE

The launch of the wave was routine, with only a few minor aircraft malfunctions reported in the strike force. In the refueling area the crews encountered clouds that reduced the visibility at times to less than a mile. Despite this handicap, all aircraft received their briefed fuel offload.

Prior to all the refuelings, to insure safe separation between cells, routes were constructed to spread the strike force out by a series of timing triangles. Then, to achieve the desired effect of maximum bombs on target in minimum time, it was subsequently necessary to compress the force by reducing the normal cell spacing. This was accomplished by having the aircraft use the compression boxes, one of which is typified by the maneuver area depicted on page 68.

Just after the post-refueling compression maneuver was complete, the #2 aircraft in the lead cell reported he had lost an engine and was having difficulty maintaining airspeed and altitude due to his heavy gross weight. Since "press-on" rules applied, the loss of the engine was not a valid reason for abort. There were two ways to handle the problem. The first option would be to let #2 fall back to the end of the formation. In the time it would take for the other cells to overtake him, he should have burned sufficient fuel so that he would be light enough to keep up with that cell. However, this would deny his original cellmates one-third of their mutual ECM support. The second option would be to have him increase power above the briefed settings. This would cause the aircraft to burn excessive fuel, depleting limited fuel reserves, thus being forced to either recover at U-Tapao or get additional fuel in a post-strike refueling. Option two was directed and post-strike refueling was arranged.

Smoke from water injection trails from the flexed wings of a B-52G. The shorter vertical stabilizer of the G model is clearly shown in the foreground.

67

LINEBACKER II | A VIEW FROM THE ROCK

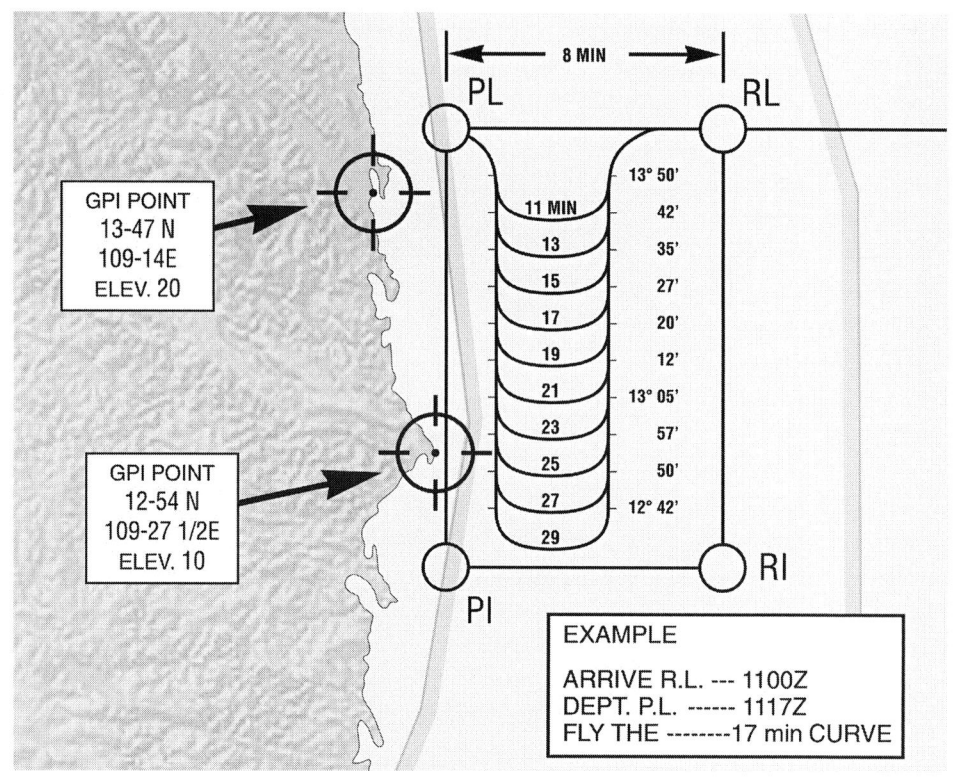

This shows the timing box located off the coast of South Vietnam, used by ARC LIGHT and LINEBACKER II forces enroute from Guam.

The inflight mission tapes described earlier captured much more than electronic signals or a mere sequence of events. In the space of a few minutes—sometimes seconds, they preserved a broad range of emotions and actions, partially revealing both what it was like to be there and some of the interplay used by the crews to cope with it. The following excerpts serve as examples. They were made by Capt Charles Core's crew from Westover Air Force Base, Massachusetts, flying as Rose 3 in Wave I, with additional radio transmissions from Rose 1 and the aircraft orbiting over the Gulf of Tonkin to monitor SAM firings. Enroute to the Kinh No Railroad Yards, and only 15 minutes from target, the tension was relieved thus:

COPILOT:	Well, there goes our nice clear night.
PILOT:	But it's undercast for us.
COPILOT:	Rog. No biggy. You don't get to see much.
NAV:	But how do you like the suspense?
PILOT:	Pretty good.

MONITOR:	SAM threat! SAM threat! Vicinity Hanoi.
NAV:	A taste of things to come.

Several minutes later, nearing the IP, as the activity directed at White and Amber Cells just ahead became intense:

PILOT:	It's triple A.
MONITOR:	SAM, SAM, vicinity Hanoi!
PILOT:	Boy, they're really puttin' it up!
COPILOT:	Chesus! They're doin' a good job of it.
NAV:	I don't blame 'em.
COPILOT:	I feel sorry for the guys down there with it.
NAV:	I feel sorry for us, too.
COPILOT:	Eh, that's true.
PILOT:	OK guys, cut out the chatter.

On the bomb run, the tension-relievers were missing, for good cause:

RADAR:	OK pilot, I'm on the target.
NAV:	That checks.
PILOT:	I've got a SAM!
EW:	And EW has an UPLINK.
ROSE 1:	Rose, UPLINK, Rose, UPLINK!
PILOT:	OK—you have an UPLINK?
EW:	That's right—OK, he's back down.
PILOT:	Got another SAM.
COPILOT:	SAM—have a visual SAM!
PILOT:	Holy Christ.
EW:	Four o'clock, comin' at us.
COPILOT:	Rose 3 has visual SAM.
ROSE 1:	SAM UPLINK, one o'clock.
MONITOR:	SAM launch, SAM launch, Hanoi!
PILOT & COPILOT:	Here come some more—could be SHRIKES—keep 'em in sight—Copilot Three, two visuals.
MONITOR:	SAM launch, SAM launch, vicinity Hanoi![45]

After that barrage of SAMs had missed, it got noticeably quieter for the crew during the last two minutes of the bomb run, only to pick up with more of the same in the PTT. This experience by Rose cell was shared by many others, where intense activity would be directed at the cell several minutes out, followed by SAM salvoes after bomb release. However, other cells experienced unremitting activity at all times on the run. These variations in experiences suggest different techniques employed by the individual sites, although aircraft position in relation to the site also determined defensive reaction capability.[46]

Colonel McCarthy, two cells ahead of Rose in the lead White aircraft, recapped what it was like for the entire wave:

The frag order called for our fighter escort to rendezvous with us approximately 15 minutes prior to the IP for the bomb run. Due to weather in their own refueling area, they were delayed so that they were unable to join up until just prior to the IP. This complicated communications and aircraft positioning, but they got there. As we turned over the IP, the EW detected the first SAM lock-on. At this time, we were northwest of Hanoi, heading for the city in a southeasterly direction. Suddenly, the gunner broke in on interphone to report that he had two SAMs low heading right for us. The EW confirmed that they were tracking toward us. The copilot then reported four missiles coming our way on the right side. Added to the pyrotechnics were SHRIKE anti-radiation missile firings, which would give us momentary concern until we identified them. Then it was nice to watch something bright streaking the other way—our guys had their own bag of tricks. The pilot then reported two missiles coming up on his side. The nav team downstairs was busy trying to complete their checklist for the bomb run. Other aircraft and Red Crown were also calling SAM warnings and antiaircraft fire. The primary radio frequency quickly became saturated. As we approached Hanoi, we could see other SAMs being fired. As one was fired on the ground, an area about the size of a city block would be lit up by the flash. It looked as if a whole city block had suddenly caught fire. This area was magnified by the light cloud undercast over Hanoi at the time. As the missile broke through the clouds, the large lighted area was replaced by a ring of silver fire that appeared to be the size of a basketball. This was the exhaust of the rocket motor that would grow brighter as the missile approached the aircraft. The rocket exhaust of a missile that was fired at you from the front quarter would take on the appearance of a lighted silver doughnut. Some crews nicknamed them the "deadly doughnuts."

The silver doughnuts that maintained their shape and same relative position on the cockpit windows were the ones you worried about the most, because that meant they were tracking your aircraft. Three of the SAMs exploded at our altitude, but in this case were too far away to cause any damage. Two others passed close to the aircraft, but exploded above us.

At about 120 seconds prior to bombs away, the SAMs were replaced by AAA fire. There were two types. Of most concern were the large ugly black explosions that came from the big 100 mm guns. Then there would be smaller multicolored flak at lower altitudes, almost a pleasure to watch by contrast. There would be a silver-colored explosion, followed by several orange explosions clustered around the first silver burst. About 60 seconds before bombs away, the flak was again replaced by SAMs. This time there were more of them and they exploded closer to the aircraft. There was no doubt about it—they were getting our range. (Note: this activity against White cell could easily explain the less intense action by the sites against Rose cell only a few minutes later.)

About this time there was a call from a cell back in the wave reporting MIGs and requesting MIG CAP. At one minute prior to bombs away, the EW's scope became saturated with strong SAM lock-on signals. It was also at this point in the run that the bomb bay doors were opened. There had been, and would continue to be, quite a bit of discussion by the staff and crews as to whether opening of the bomb doors, exposing the mass of bombs to reflect radar energy to the SAM sites, gave the enemy an even brighter target to shoot at.

About ten seconds prior to bombs away, when the EW was reporting the strongest signals, we observed a SHRIKE being fired, low and forward of our nose. Five seconds later, several SAM signals dropped off the air, and the EW reported they were no longer a threat to our aircraft.

The BUFF began a slight shudder as the bombs left the racks. The aircraft, being relieved of nearly 22 tons of ordnance, wanted to raise rapidly, and it took a double handful of stick and throttles to keep it straight and level. After the release was complete and the bomb doors closed, Tom Lehar put the aircraft in a steep turn to the right. A second later, a SAM exploded where the right wing had been. The turn had saved us, but the gunner and copilot reported more SAMs on the way.

It seemed like the turn was going to last forever, and the copilot reported the SAMs tracking the aircraft and getting closer. It was now a race to see if we could complete the turn before the SAMs reached our altitude. Once the turn was completed we would be free to make small maneuvers, because the other aircraft in our cell would still be in the . . . turn. . . .

It is strange what goes through your mind at a time like this. My thoughts were: "What the hell am I doing here?" With a lung full of what eventually turned into double pneumonia and no ejection seat I wasn't exactly an ideal insurance risk.

We completed the turn and started our maneuver as the SAMs reached our altitude, but they did not explode. One on the left and another on the right seemed to form an arch over the aircraft. As they approached each other at the apex of the arch they exploded.

We saw another SHRIKE missile launched, and again some of the threat signals disappeared. Those F-105 Wild Weasel troops from the 388th Tactical Fighter Wing were earning their pay tonight. I made a mental note to write their wing commander, Colonel Mele Vojvodich, and congratulate his crews on their outstanding work. Mele and I had served together at Korat.

As we egressed the target area the SAM signals dropped off. Now our problem was to get the wave back together and assess our battle damage, plus collect strike reports. It was the *duty of the ABC to collect this information and then pass it through encoded HF radio communications to the 8AF command post. If any aircraft were damaged, it was also up to the ABC to make the decision as to where the aircraft would go. If it was only minor battle damage, then the aircraft would normally be instructed to return to Andersen with the rest of the bomber stream. We needed every airframe we could get our hands on for the following day's sorties.*[47]

In the event an aircraft was severely damaged, it would usually head for U-Tapao, where heavy maintenance repair capability was available. Should the damage be so bad that the aircraft couldn't make it to U-T, the crew would be instructed to head for a closer base in Thailand, or a long runway such as at Da Nang.

Some of these emergency landings had become sporty propositions. An aircraft hit before LINEBACKER II began gave a typical example of the outstanding flying skills of some relatively inexperienced pilots.

Hit by a SAM over North Vietnam the previous April, the pilot reported rapid loss of fuel, two engines completely out, and the adjoining two with only partial power. All fuel gauges were spinning, making it impossible to accurately determine the amount of fuel on board. The SAM had exploded close enough to the aircraft to blow off part of the tip tank and put some 400 holes throughout the airframe.[48] It also knocked out some of the flight instruments and the cockpit lights. The pilot was instructed to try to make an emergency landing at Da Nang, since it was the closest available airfield that could handle B-52s.[49]

When the aircraft arrived over Da Nang it was nighttime. The field was under instrument flight conditions (IFR)—that is, not visible from altitude—and was additionally undergoing a rocket and mortar attack. At each end of the runway there were mine fields, and Viet Cong snipers with automatic weapons that previously had done their share of damage to U. S. planes.

The aircraft commander, John Alward of Robins Air Force Base, Georgia, a captain with 1200 hours total flying time, had the option of bailing his crew out and heading the aircraft out to sea, or attempting a landing at Da Nang. He elected to attempt the landing, and his crew chose to stick with him. As he started his instrument approach, the base took an increasing number of incoming rounds.

The approach and landing speeds of the BUFF vary considerably with the fuel on the aircraft at the time. Because of the spinning gauges and the massive fuel leaks, the crew could only estimate the fuel load. They properly estimated on the high side to keep from stalling out and, as a result, when Capt Alward flared for landing he was "hot" and landed long. Then, when Capt Bob Davis, his copilot, pulled the handle to put out the large drag parachute to decelerate the aircraft, nothing happened. It was later determined that a SAM fragment had severed the drag chute actuator cable.

CHAPTER 3 | ACT ONE

Faced with the upcoming mine field, the pilot elected to make a go around. Handling a BUFF with two engines out on one side is considered a very serious emergency. One with two engines out and other battle damage, at night, in weather, with part of the flight instruments out and fuel spraying out of body tanks is considered difficult even for a highly qualified instructor pilot to handle.

Captain Alward wasn't concerned with the impossibility of the situation. Using every bit of the long runway, he pulled his damaged aircraft into the air just prior to the wheels going onto the overrun and the mine field.

On the next approach, he was able to make a successful landing. Although severely damaged, the aircraft was eventually flown back to U-Tapao and repairs were completed.

But, that was in the past. Back over North Vietnam, there was work to be done. All aircraft in Wave I reported in, and the news was good. It had been "hairy," but all had made it, with only one aircraft indicating some possible battle damage. All of that plane's systems were still operating, so they were directed to return to Andersen with the rest of the wave. Using the lesson learned from the first night, larger pre-target air refueling onloads had been arranged for the second night, and future cells would not normally need a post-strike refueling.

A significant change in tactics was permitted after Wave I had struck its targets. The word was flashed, and subsequent aircraft were allowed to perform the TTR maneuver from IP to target, provided they maintained cell formation and were straight and level prior to bombs away.[50] This was made possible because radar camera film from the first night's raids had been developed and analyzed, revealing that all bomb aiming points showed as predicted. Thus, the RN and Nav did not need as much time to positively identify the aiming points and could readily make last-minute refinements. Enough straight and level time was allowed for the bombing system gyros to stabilize for an accurate bombing platform.

Wave I's good fortune was not completely shared by Wave II. Hazel 3, in the lead cell flying directly toward Hanoi from the west northwest, received minor damage, but completed the mission and returned to Guam. This, by the way, was the only G model to sustain damage and not be forced down during the entire campaign.[51] Three cells behind Hazel, Capt John Dalton's Westover crew in Ivory 1, a D model, got it much worse. However, they managed to limp back to a successful, but tense, landing at the Marine base at Nam Phong, Thailand.[52]

Wave III struck the Thai Nguyen Thermal Power Plant and the Yen Vien Complex. There were numerous SAM firings against the nine cells going against Thai Nguyen, but none seriously threatened the force. SAM reaction to the attack by three cells on Yen Vien was sporadic; Wave III came through unscathed. The entire night's force of attackers had made it without the loss of a single aircraft.[53]

This situation was a critical determinant in the overall plan. After the losses on Day One, which were expected, the force had turned right around and gone with the same game plan on the second day—with no losses. One must appreciate how much of a tempering effect this had on the decision-making process at all levels. All three waves had driven to within ten miles or less of the capital city with part or all of their cells and emerged to fly again. It did not go unnoticed that the Yen Vien raiders of Wave III had gone in five hours after the attack by Wave II, and had experienced remarkably light retaliation.[54]

Partly because there were no losses, and because of long lead time from planning to execution, SAC Headquarters made the decision to continue with the same attack plan on December 20th. Yen Vien would be attacked on an identical course that night, at which time the "bubble would burst" unmercifully. But, at this point in the campaign, the indicators vindicated the battle plan.[55]

The hydraulic ram force of SAM fragments, with an estimated penetrating capability at 75 feet at least equal to a .50 caliber all-purpose round at muzzle velocity, is evident on this upper wing surface.

CHAPTER 3 | ACT ONE

Weary crewmembers reconstruct their missions for the debriefers, following flights of up to 18 hours duration.

As the last G model from Cinnamon Cell was touching down at Andersen in the early afternoon of December 20th, Quilt Cell was starting engines. Perhaps, as the weary crews climbed out of their planes, they saw Capt Terry Geloneck taxiing Quilt 3 into the lineup. The Rock would never feel the weight of his G model again, and the Hanoi Hilton would be gaining some new residents, this time from Beale Air Force Base, California. Still, the glare of the tropic sun on exhausted eyes probably made all of the flightline activity seem more like a transient dream than anything to pay attention to, and the last of Day Two's successful flyers headed for debriefing.

DAY THREE—THE DARKEST HOUR

The frags were later than normal coming in for Day Three, because of last minute changes to targets, tactics, and assessments of the enemy air order of battle. There was considerable discussion between the staffs and crews on the desirability of post-target turns (PTTs) after bomb release, and on other tactics. These concerns were expressed to SAC Headquarters through 8 AF.

Lt Gen Glen W. Martin, Vice Commander-in-Chief, SAC, experienced that concern first-hand:

> We in SAC perceived early in LINEBACKER II that mini-changes in timing and in aircraft course, speed, and altitude could make the difference between a hit and a miss in the daily battle between the B-52s and the communist SAMs.

The same considerations applied to the joint operations which included not only coordinated strike operations by the 7th Air Force and 7th Fleet, but diversionary, ECM electronic and chaff dropping, and MIG CAP operations as well. As a consequence, the operational planning and tactical analysis were incredibly complex, with very little time available in the 24-hour kaleidoscope to take advantage of all crossfeed and potential improvements.

Nor was that all. We also learned early that the B-52 ECM systems needed immediate adjustment for better protection against enemy SAM guidance radar. At nearly the speed of light, the Air Force Systems Command set up a series of tests involving SAC B-52s flying against enemy radar (basically Soviet). SAC people spent sleepless nights with Systems Command people at Eglin, Florida, during LINEBACKER II, finding the best ways to block out the SAMs. And, I'm proud to say, U.S. industry responded successfully on a minute-to-minute basis, with needed alterations.[56]

The force of the explosions blew huge sections of railroad tracks completely off their beds.

CHAPTER 3 | ACT ONE

Real-time testing lagged the missions themselves, of necessity, and the third day's tactics and maneuvers remained basically the same as those on the 19th.[57] These maneuvers, which by now included the highly advantageous pre-target evasive tactic, were supplemented by instructions to the Electronic Warfare Officers (EWs) to 'revise the procedures for jamming the SA-2 system.[58] Short of that, Day Three's mission could best be described as a composite of routes, targets, and tactics from Days One and Two.

By the time the first cells started launching from the Rock on the 20th, the base's learning curve had progressed rapidly. The support problems encountered on the first day had largely been eliminated, and Andersen's activities were running more smoothly. The only problem supervisors reported was that they could not get their troops to go home after they finished a work shift. These men wanted to be there for the launch of their aircraft. Never had such large numbers of Stratofortresses been launched at one time on a combat mission. These men were helping to write a chapter in the history of air power, and they didn't want to miss it.

The around-the-clock operations of the first three days found aircraft awaiting takeoff as others completed their mission.

Unknown to them then, word had already come down extending the three-day maximum effort to an "indefinite" status.[59] Therefore, their constant straining to support the surge would know no relief. Officer and airman alike, when told that tomorrow would be more of the same—day after day—just shrugged. And on the morrow they did it again—day after day.

LINEBACKER II | A VIEW FROM THE ROCK

Crowds of "Red Ball" specialists, prepared for last-minute maintenance on a launching bomber, watch a G model land from the previous night's raid.

December 20th witnessed the most pronounced NVN defensive effort against the B-52s, and the highest single-day losses of LINEBACKER II.[60] Several cells reported MIG engagements, while others had indications that the MIGs were flying parallel courses to the attackers, thus relaying their altitude and airspeed to the ground defenses.[61] Antiaircraft artillery was again heavy, but usually short of the mark.[62] Belying indications from the previous night that SAM supplies might be getting low, or that the defensive net was breaking down, the enemy hurled over 220 SAMs at the three waves that night. Volume of fire was accompanied by discriminatory firing patterns. The NVN at times did not engage the first cell over target, but used it to determine flight paths and the turning points.[63] Subsequent cells then experienced intensive salvoes near the release points, where they were committed to stabilized flight, and in the post target turn.[64]

It has been speculated that the intensive reaction to the third day was no mere coincidence, but was tied to a perception of our original intentions—a three-day maximum effort. It would have been useful for the NVN cause if this "last" max effort could have been repulsed. Fortunately, it didn't work that way.

All attacks on Hanoi were again from a narrow wedge out of the northwest, even narrower than on the previous two days. Of the eleven cells which made up Wave I, nine were tasked

CHAPTER 3 | ACT ONE

against the Yen Vien Railroad Yards and adjacent Ai Mo Warehouse Area. At Yen Vien there had been light retaliation the night before.[65]

But this night all hell broke loose. Quilt Cell, with two ECM degraded aircraft, led the attack. Quilt 3 was knocked down in the post-target turn.[66] Gold and Wine made it. Next came Brass. Brass 2, also with degraded ECM, was hit in the PTT. Seriously crippled, the aircraft made it back to Thailand before the Loring Air Force Base, Maine, crew had to abandon it.[67] Now it was the D models' turn. Snow and Grape Cells got through, followed by Orange. Orange 3 was hit by two SAMs only seconds prior to bomb release and exploded.[68] Four of the Westover Air Force Base crew, commanded by Maj John Stuart, became MIA.

Captain Rolland A. Scott, TDY to the 72d SW from Barksdale Air Force Base, Louisiana, had the unpleasant experience of sampling everything that went on in Wave I, but survived:

I flew in Gold 2 with another crew as a substitute pilot. The time, track, and target location were nearly the same as my mission on the 18th. Shortly after takeoff, we lost one engine and flew the mission on seven. That wasn't too serious a problem in the "G" model, but I would have felt better if it hadn't happened.

On the northbound leg over NVN we heard a good deal of fighter activity and numerous sightings were made of aircraft with lights on, presumably friendly fighters. There appeared to be no SAM activity.

On the southeast leg approaching the IP, 'my copilot stated he saw a MIG21 on the right wing of our aircraft. In mild disbelief, I stretched to see out his window and sure enough, a MIG-21 with lights off was flying tight formation with us. I believe we could actually see the pilot. The approach of the fighter had not been detected by onboard systems. Shortly, two or three minutes, the copilot reported the MIG had departed. Almost immediately I saw the same, or another, enemy aircraft flying formation on the left side of us. After a brief period, less than a minute, it departed.

Our sighs of relief were short-lived, and we quickly learned what the MIGs had been up to. We visually detected missiles approaching from our eleven and one o'clock positions. Several pairs of missiles were simultaneously launched from these directions. I was extremely worried that missiles were also approaching from our rear that we could not see. The EW reported no UPLINK or DOWNLINK signals with the missiles this mission as were reported on the night of the 18th. However, these missiles appeared to be a lot more accurate than on the 18th. They seemed to readjust their track as I made small turns. I waited for each to get as close as I dared, and then would make a hard, although relatively small, maneuver in hopes of avoiding them.

They arrived in pairs, just a few seconds apart. Some, as they passed, would explode—a few close enough to shake my aircraft. In fact, one exploded so close and caused such a loud noise and violent shock that I stated to the crew that I thought we had been hit. In a very few seconds, after assessing engine instruments and control responses, and having received an OK from downstairs, I determined we had not been hit, or were at least under normal control, and we continued the bomb run. Apparently the MIG-21 we saw was flying with us to report heading, altitude, and airspeed to the missile sites.

The missiles were no longer directed toward us in the latter half of the bomb run; however, I could see SAM activity ahead in the vicinity of the target. In fact, while on the run we saw a large ball of fire erupt some few miles ahead of us and slowly turn to the right and descend. I thought it was a BUFF and was sure no one would survive what was apparently a direct hit. I later learned what I saw was Quilt 3 going down in flames. Amazingly, four crewmembers successfully ejected.

We completed our bomb run and were in the middle of our post-target right turn when we again became an item of interest to the missiles. From our left and below were at least three missiles, perhaps four, approaching rapidly. I felt I had no chance to avoid them by either maintaining or rolling out of the right turn, so I increased the planned bank angle drastically ... and lowered the nose. The SAMs passed above us from our left. I lost some altitude in the maneuver, and in the attempt to climb and accelerate on seven engines I lagged behind lead and somewhat out of position.

There were no further SAMs directed at our aircraft; however, there was apparently a lot of enemy fighter activity on our withdrawal, according to radio transmissions. We could see numerous fighters with lights on, and the gunner reported numerous targets on radar, one of which appeared to follow us, but not in the cone of fire. We saw no aircraft which appeared to be hostile, nor any hostile maneuvers.

As we passed east of NKP on a southerly heading, we heard what was apparently a B-52 crew abandoning their aircraft over friendly territory. In the distance, toward NKP, we soon saw a fireball which we assumed to be a BUFF impacting the ground. It must have been Brass 2, and Capt John Ellinger and his crew were mighty lucky.[69]

Two Gs and one D were down, and three of nine cells had experienced losses. The optimism that had accompanied Day Two's results was shattered. The shock waves went half-way around the world.

At SAC Headquarters, Omaha, Nebraska, the Deputy Chief of Staff for Intelligence, Brigadier General Harry N. Cordes, watched it unfold:

The SAC staff virtually lived together throughout LINEBACKER II. No one slept very much. The days and nights were spent together, selecting targets, developing tactics, explaining results.

CHAPTER 3 | ACT ONE

Therefore, the Darkest Hour was just as dark at SAC Headquarters as it was at Guam and U-Tapao. In addition to the personal concern of the SAC staff, there was pressure from many external sources. Such expressions as "stop the carnage—we can't lose any more B-52s—it has become a blood bath" were commonplace. General Meyer, Commander-in-Chief, SAC felt many pressures. Many people in Washington were worried that the Air Force would fail—that the U.S. couldn't bring Hanoi to its knees. Many senior Air Force people were concerned that if the bombing continued, we would lose too many bombers and airpower doctrine would have proven fallacious. Or, if the bombing were stopped, the same thing would occur. Admiral Moorer, Chairman JCS, was concerned, but left the ultimate decision to General Meyer.

Midway in the disastrous third day, when loss reports were coming in, and the next wave was nearing the recall line, General Meyer assembled the staff. We reviewed with him every single aspect of the situation:

- *Damage inflicted on the enemy*
- *Aircraft loss rates, damage rates*
- *Apparent causes of losses and damage*
- *Enemy defense status-running low on SAMs?*
- *Current tactics, new tactics, ECM*
- *Support packages and tactics*
- *Possible new targets outside high threat areas*
- *Crew morale and discipline*
- *Penetration analysis loss predictions*
- *Air Force doctrine and history:*
  - *No defense against a determined air attack*
  - *Never turned back due to enemy action*

General Meyer experienced first-hand the "loneliness of command." He and he alone must make the decision. He listened as a judge to all the evidence. He polled every single man in the room—general, colonel, captain, lieutenant—go or no-go? He polled Jerry Johnson—can the crews take it? Then he made his decision, probably the most difficult of his career—

"PRESS ON!"[70]

That monumental decision by the CINCSAC measured just how closely the scales had come to being tipped in an unknown direction. In making it, he was not giving blanket approval to a mindless continuation of things as they had been. On the contrary, everything was being evaluated on a realtime basis. An example of this was seen at the very time the command was directed to press the attack, in two cells of G models already enroute to Hanoi that were recalled from Wave II.[71]

81

Losses, damage, and near-misses had revealed a pattern. Of the G models deployed to Andersen, roughly half had been modified with a more extensive ECM package than had the other half. It was now apparent that the unmodified Gs were neither protecting themselves nor their formation adequately, and were bearing the brunt of the losses inflicted by Hanoi's SAM sites.[72]

It was determined that the six recalled Gs in Wave II would not significantly alter the tonnage delivered on the targets they were fragged against, so the wave drove on without them. The bulk of the wave's sorties were targeted against the Thai Nguyen Thermal Power Plant for the second day in a row, with two cells dropping on the Bae Giang Transshipment Point. Both targets lay well north and northeast of Hanoi. Coming in four hours behind Wave I, this wave experienced no losses or battle damage.[73]

Wave III, with three of its four targets in and around Hanoi, turned into the second half of the nightmare. The bomber fleet contained both modified and unmodified Gs, and the SAC staff knew it. However, the target priorities and weight of effort dictated that all aircraft be committed to the assault. Therein was found part of the conditions which had made the CINCSAC's decision so difficult. Twelve Gs were to attack the Kinh No Complex, an area so extensive that it contained four distinct targets. To also withdraw them from the strike would not only cause an unacceptable shortfall in the desired level of destruction in this lucrative target but, when added to the six Gs already withdrawn, would mean that over one-half of the combined effort of Waves II and III against Hanoi would be deleted. They pressed the attack.[74]

Only minutes ahead of the Gs, nine D models struck the Hanoi Railroad Repair Shops at Gia Lam. Straw 2, the fifth aircraft in, took a SAM hit in its PTT from a missile most probably fired by site VN 549.[75] The aircraft made it as far as northern Laos, where all crewmembers, except the radar navigator, were recovered by HH-53s of the 40th Aerospace Rescue and Recovery Squadron. The first D model from the Rock had gone down.[76]

Eight minutes behind Straw, Olive 1 led the attack on Kinh No. An unmodified G, it took a hit after the release.[77] Of the crew from Fairchild Air Force Base, Washington, only Lt Col Jim Nagahiro, the pilot, and Capt Lynn Beens, the navigator, were returned POWs. Lt Col Keith Heggen, the Deputy ABC and seventh man aboard, died in prison from his battle wounds.[78]

Two cells back from Olive was Tan 3. It was an unmodified G that had additionally lost its bombing and navigation radar. Such a problem was compensated for by a tactic whereby the gunner in the aircraft ahead used his radar to relay precise range and bearing data to the affected aircraft. They could then compute a very accurate release point of their own from this information. While trying to follow the directions of their cellmate to fly to the target,

CHAPTER 3 | ACT ONE

the Blytheville Air Force Base, Arkansas, crew became well-separated from the other two aircraft. As Tan 1 and 2 approached the release point, a SAM exploded under Tan 3, and it went into a dive. Captain Randall Craddock struggled to get the aircraft back on altitude and course. Shortly thereafter, another SAM impacted the isolated aircraft and the pilot ordered bail-out.[79] The gunner, SSgt Jim Lollar, escaped from the aircraft before it exploded and was the only one of the crew to become a POW, the rest being declared KIA or MIA.[80]

Captain Chris Quill and his crew from Barksdale Air Force Base, Louisiana, were driving in on the same route as Olive and Tan against Kinh No. Their cell, Aqua, would be the last three G models to "go downtown" during the campaign. Although they themselves were the object of intense attack, they were eyewitnesses to the grim larger pattern unfolding in front of them. Major Dick Parrish, the RN, clearly recalls those events:

As we made our turn north of Thud Ridge, both the pilot and copilot saw a burning aircraft at a lower altitude heading back to the northwest. We were an item of interest to the SAMs at the time, and were having problems enough of our own. However, they were able to look at the burning plane long enough to be satisfied that it was a BUFF. [Note: this was most probably a sighting of Straw 2, the D model which made it back to Laos.]

As we pressed on, I heard Chris and Joe Grinder, our copilot, exchange the following remarks: "Good Lord, what was that?" "Must have been a direct hit." "My God, what a fireball!" [Note: Due to the violence of the explosion, this was most probably Tan 3.] Right after that, the EW yanked us back to our own situation by stating that his scope was covered with threat signals.

I couldn't worry with the EW's threats, or fireballs, or anything else. I had only one job—to get the bombs on the target with no mistakes. I had half-seriously told Bill Stillwell earlier that if he sat over there in the navigator's position and let me forget to open the doors, I'd kill him. He didn't forget—and neither did I. I wasn't about to be on a crew of six people, going all the way up there and risking our lives, only to not get the bombs out. We made it through release and the big turn back to the west. After we had rolled out, Chris was considering putting the aircraft back on autopilot, because it had become so quiet around us. This procedure would allow both pilots more time to concentrate outside the aircraft. Both he and Joe decided to take one more good look beforehand, though. As Chris looked as far back and down to the left as he could, he spotted two white streaks coming at us. The next thing I knew, we were in a steep, descending right turn. Almost instantly, Leo Languirand saw two traces come onto his gunnery scope. While we continued to maneuver, the traces continued to climb, and were closing. Then the two blips disappeared. We did a little mental gymnastics and figured they went off just about where we would have been.

It got quieter as we headed for Laos. Then, as we were nearing the border, Chris and Joe saw a large explosion on the ground out ahead of us. If it was an airplane, which they were sure it was, it had to have been a big one. (Note: this could have very possibly been another sighting of Straw 2.)

All of the sightings the pilots had made, plus what we had experienced on our own sort of got to us, I think. As we headed on down-country and towards the water, I tried to break the ice with some weak joke. Dick Engkjer, our EW, had been staring at all that wild stuff on his scope, and he didn't think I was very funny. He promptly chewed on me for trying to act happy at a time like that. I understood how he was feeling. But then I got to thinking, and said, "What the heck. We did the job and we're out in one piece. I think there's plenty to be happy about." The flight back home went better for me after that.[81]

A scene of total destruction in the Ai Mo Warehouse Area.

Brick Cell, the last ones into the immediate Hanoi area that third night, attacked the Hanoi Petroleum Products Storage Area. The large number of SAMs fired at Brick Cell after bomb release nearly proved to be the undoing of Brick 2. However, a proximity SAM detonation was not enough to bring the aircraft down, and it safely recovered at U-Tapao.[82]

A battered Wave III headed for the long flight home. As far as attacks were concerned, Day Three was over. The overall box score was grim: four Gs and two Ds knocked down, with a third D damaged.[83] All the lost Gs were unmodified; four losses and the one battle damage occurred after bomb release. A new battle plan had to be developed if the Stratoforts were to continue their attacks in the Hanoi area.[84]

CHAPTER 3 | ACT ONE

NOTES

1. *43SW History*, p. 64.
2. *8AF History, V II*, p. 342.
3. *43SW History*, pp. 63-64.
4. *Ibid.*, p. 64.
5. Colonel Hendsley R. Conner, Narrative written to authors, 13 May 1977.
6. Message (TS), JCS to CINCPAC, JCS 5829, for Gayler from Moorer, 18/00152 December 1972 (72-B-7620). TOP SECRET. Subsequently declassified on 16 December 1977.
7. *8AF History, V. II*, p. 352.
8. *Chronology*, pp. 27 and 59.
9. *43SW History*, pp. 73-74.
10. *8AF History, V. II*, p. 377.
11. *Chronology*, p. 53.
12. *303CAMW History*, p. 24.
13. *43SW History*, p. 62.
14. *Ibid.*, pp. 69, 72, and 104.
15. *USAF AIROPS*, p. IV-287.
16. *Ibid.*, p. IV-171.
17. *43SW History*, p. 104.
18. *Chronology*, p. 60.
19. *SAC Participation*, p. G-12.
20. *8AF History, V. II*, p. 358.
21. *8AF History, V I*, p. 66.
22. *43SW History*, p. 25.
23. Notes by Author McCarthy, Blytheville AFB, AR., autumn 1977.
24. *Flight Manual(s)*, T.O. IB-52D-l, l August 1974, and T.O. IB-52G-I. 1 January 1975, Air Logistics Command, Tinker AFB, OK, Excerpts.
25. *Chronology*, p. 62.
26. *Ibid.*, pp. 60-61.
27. *USAF AIROPS*, pp. IV-291 to IV-292.
28. *Ibid.*, p. IV-278.
29. *Ibid.*, p. IV-150.
30. *Ibid.*, p. IV-312.
31. Notes by Author Allison, Blytheville AFB, AR., autumn 1977.
32. Colonel Hendsley R. Conner, *op. cit.*
33. *The USAF in Southeast Asia 1970 1973, Lessons Learned and Recommendations: A Compendium*, CORONA HARVEST, Prepared by HQ PACAF, Hickam AFB, HI, 16 June 1975, pp. 110-111. Hereafter cited as *USAF in SEA*. SECRET
34. *Chronology*, p. 69.
35. Chaplain (Capt) Robert G. Certain, Conversation with Lt Col Allison, 24 May 1977.
36. *Damage Analysis*, p. A-67.
37. *The New York Times*, 24 December 1972, p. 3.
38. *Damage Analysis*, pp. A-121 and A-137.
39. *Chronology*, pp. 95-96.
40. *History of 307th Strategic Wing, October-December 1972*, Volume I, U-Tapao Royal Thai Navy Airfield, Thailand, 12 July 1973, p. 97 and Exhibit 30. Hereafter cited as *307SW History*. SECRET
41. *Chronology*, pp. 95-96.
42. *SAC Operations in LINEBACKER II, Tactics and Analysis*, briefing prepared by HQ SAC/XOO, Offutt AFB, NE, 3 August 1976, presented to authors in April and June 1977. Hereafter cited as *SAC OPS*. SECRET

43 *8AF History, V. II,* pp. 508-509, 517.
44 *43SW History,* p. 26. See also *303CAMW History, op. cit.,* p. 18.
45 Combat Mission Tape, Westover Crew E-12, 19 December 1972, cassette tape on file at A.F. Simpson Historical Research Center, Maxwell AFB, AL.
46 *SAC Participation,* pp. K-1 to K-3.
47 Recollections, Blytheville AFB, AR., winter 1977. For supporting data on 100 millimeter antiaircraft artillery, see also *COMMANDO HUNT V,* Report prepared by HQ 7AF, Tan Son Nhut AB, South Vietnam, May 1971, p. 239. SECRET
48 *Damage Analysis,* pp. A-117 to A-118.
49 *History of 307th Strategic Wing,* April-June 1972, Volume I, U-Tapao Royal Thai Navy Airfield, Thailand, 25 September 1972, p. 101.
50 *Summary of Tactics,* Report prepared by 8AF /IN, Andersen AFB, M.I., 1 February 1973. SECRET
51 *Damage Analysis,* p. A-141. The document errs on the date of this incident.
52 *Chronology,* p. 106.
53 *Ibid.,* pp. 109 and 11 I.
54 *SAC OPS, op. cit.*
55 *Ibid.*
56 Lt General Glen W. Martin, narrative written to authors, 15 August 1977.
57 *Chronology,* p. 121.
58 *SAC Participation,* p. K-3. See also *8AF History, V. II,* p. 362.
59 Message (TS), JCS to CINCPAC, CINCSAC, and COMUSMACV, JCS 7807, for Gayler, Meyer, and Weyand from Moorer, 19/2322Z December 1972 (72-B-7673). TOP SECRET, subsequently declassified on 16 December 1977.
60 *Chronology,* pp. 143-144. See also *43SW History,* p. 76.
61 *8AF History, V. II,* p. 365.
62 *COMMANDO HUNT VII,* report prepared by HQ 7AF, Tan Son Nhut AB, South Vietnam, June 1972, p. 232. SECRET
63 *43SW History,* pp. 138-139.
64 *USAF AIROPS,* p. IV-288.
65 *Chronology,* p. 125.
66 *Damage Analysis,* pp. A-79 to A-80.
67 *Ibid.,* pp. A-63 to A-64.
68 *Chronology,* pp. 126-127. See also *8AF History, V. II,* pp. 365-366.
69 Maj Rolland A. Scott, narrative written to authors, 21 October 1977.
70 Brig General Harry N. Cordes, narrative written to authors, July 1977.
71 *Chronology,* p. 131.
72 *USAF AIROPS,* p. IV-287.
73 *Chronology,* pp. 132-133.
74 *SAC OPS.*
75 *Damage Analysis,* p. A-60.
76 *Chronology,* p. 135.
77 *Damage Analysis,* p. A-71.
78 *Chronology,* pp. 137-185.
79 *Damage Analysis,* pp. A-83 to A-84.
80 *Chronology,* pp. 138-139.
81 Maj Richard L. Parrish, oral narrative to Lt Col Allison, 24 October 1977.
82 *Chronology,* p. 140.
83 *43SW History,* p. 176. See also *Chronology,* p. 143.
84 *USAF in SEA,* pp. 114-115.

CHAPTER 4 | ACT TWO
DAY FOUR – THE PLOT SHIFTS

The two tactical wing commanders spent the morning of December 21st, along with their executive officers and the chaplains, personally meeting with the missing crewmembers' wives who were on Guam at the time. Wives of the permanent party officers also organized an effort to help these and other visiting women get through a very traumatic period. Their attempts to compensate for months of TDY separation by sharing a pleasant tropical Christmas with their husbands had, without any forewarning, become a period of tension or tragedy.

Day Three marked the completion of the first phase of bombings, and the weight of effort for Day Four had already been established as coming from the D fleet at U-Tapao, independent of the previous day's results. An assessment and evaluation of the sudden upsurge in losses, especially among the G fleet, could therefore be made without the added stress of having to decide on the spot what was to be done about the current day's scenario. It gave many members of the command on both sides of the world time to apply their expertise to the problems the Gs were encountering, while also zeroing in on analyses of tactics by the entire force.[1]

Andersen's respite from LINEBACKER II participation on Day Four would be repeated on Days Five and Seven as a continuing part of the overall plan for the second phase of the operation. The planned sortie rate for this phase was reduced to 30 aircraft per day, well within U-Tapao's capability to handle alone. Logistics considerations overwhelmingly favored conducting the strikes from only one base, and many benefits derived from U-T's significantly shorter mission length of approximately four hours.

Last-minute changes to targeting or tactics or both could be managed more effectively from the Thai base due to the reduced time span from launch to desired time on target (TOT). Additionally, these changes would have to be coordinated with only the one base. Air refueling was not needed. Precise navigation timing was easier to effect. Fuel weight could be traded off for bigger bomb loads. The lesser fuel requirements were also one of several considerations which provided for easier and faster turnaround after post strike at U-Tapao. Shorter route lengths meant that the aircraft were available for maintenance whereas Andersen aircraft would still have six or more hours of flying before completing their missions.² Similarly, the crew force posture benefited from the increased time available for crew rest. That, however, was a dubious compensation for being shot at on a daily basis.

During preparatory maintenance, each parking hardstand, such as this one at U-Tapao, was crowded with special equipment and weapons. The clip-in assemblies shown are loaded with 750-pounders.

Andersen committed 30 aircraft to support of the war in South Vietnam, reverting to the standard separate three-ship cyclic operations.³ This provided a chance to get some of the new crews who had just come in from the States started on their checkout programs. The judgment was that they needed to practice cell procedures on at least three missions in the South before going to downtown Hanoi. In the meantime, U-Tapao's experienced crews would be working over three new targets—Hanoi Storage/Bac Mai, Van Dien Storage Depot, and Quang Te Airfield.⁴

This turned out to be a tough night for two of the crews sent against the Hanoi Storage/Bac Mai target. Captain Pete Giroux and his crew from March Air Force Base, California, experienced failure of the bombing radar just prior to the IP. In a desperate but determined attempt to move from the lead position of Scarlet Cell to a trailing position, from which they could be guided to their target by their cellmates, they became separated from the other two aircraft. Thus isolated, they fell victim to an alert SAM team and were shot down.[5] Prior to the SAM hit they were also engaged by MIGs, and Scarlet 2 was ineffectively fired on by a MIG after the PTT. Equipment failure—cell separation—MIGs—SAMs: Scarlet had it all.[6]

Four minutes behind Scarlet, Blue Cell was bearing down on Bac Mai. The EW aboard the lead aircraft, Lt Col Bill Conlee from Carswell Air Force Base, Texas, relates his LINEBACKER II experiences, providing insight as to what it was like at both U-Tapao and Hanoi:

When LINEBACKER II kicked off on December 18th, our crew was chosen to be wave lead on the last wave of the first day from U-T. After the general briefing crews split up into specialties to discuss mission tactics, late intelligence, and bombing information. Excitement ran high and the general feeling was "At last—let's do it!"

Launch and join up was no problem and the mission went routinely until entry into North Vietnam, where a high level of radar activity was immediately apparent. Electronic countermeasures were applied. We proceeded to the IP and completed our bomb run without incident except for obvious SAM radar engagement attempts. We released our weapons and started our post release turn when SAM launches were detected, visually and electronically. From the time we started our turn until about 40 miles southwest of Hanoi, we came under heavy SAM attack, employing evasive action turns to avoid SA-2s. During this seeming eternity of time, we counted approximately 40 SAMs launched in our general direction or in the vicinity of our cell. On the return route to U-T we were unable to contact Rose 1, and someone reported they were down near Hanoi. After landing, the debriefers were incredulous at the large number of SAMs which had been fired during the mission.

On December 19th we were again selected to fly wave lead as Green 1 against the Thai Nguyen thermal power plant, 40 miles north of Hanoi. This mission went very smoothly, despite some high altitude AAA fire and sporadic SAM firings, which were wild and not effective against our wave. No losses were incurred despite SAM radar engagement.

Crew morale during this period was very high. This was brought home to us on the night of December 20th when our crew pulled #1 standby crew. This meant we waited all night in a spare B-52D for any crew to abort so that we could fly, but nobody aborted. All bomber cells launched like clockwork.

The following night, December 21st, we were scheduled to lead Blue Cell, with a call sign of Blue 1. Our target was again in Hanoi [Bac Mai Storage], with a release time of 0347 local time on the 22nd. This mission was routine until we reached the IP for our bomb run. At this time the copilot, Captain Dave Drummond, remarked: "It looks like we'll walk on

SAMs tonight," as he could see numerous SAM firings ahead of us. This comment proved only too true.

Between IP and bomb release point, ten SAMs were fired in the vicinity of Blue Cell. At bombs away we were bracketed by two SAMs, one going off below us and to the left, the second exploding above us and to the right. Shrapnel cracked the pilot's outer window glass, started fires in the left wing, and wounded Lt Col Yuill, the pilot; Lt Col Bernasconi, the radar navigator; Lt Mayall, the navigator; and myself. We also experienced a rapid decompression and loss of electrical power. Shortly after this, with the fire worsening, Lt Col Yuill gave the emergency bailout signal via the alarm light and I ejected from the aircraft.

During free-fall, two more SAMs passed me, and I attempted to look for our aircraft, but was unable to see it. I was also unable to see any other chutes in the darkness. I realized after my chute had opened at preset altitude that my left arm was numb and that I had lost my glove from my left hand during ejection. I also realized that I was bleeding profusely from the face and arm due to shrapnel wounds. I deployed the survival kit with my right hand and prepared for landing. I steered for an open field, just missing going into a large river, and believed that I would land undetected.

About 200 to 300 feet from touchdown, my illusions were shattered when small arms fire was directed at me. I ignored the firing and concentrated on making a good landing. I touched down, dumped my chute and took off my helmet, and at once was set upon by a mob of North Vietnamese, both civilian and military. They immediately took my gun, my watch, and my boots. They then stripped me at gun point to my underwear and forced me to run for approximately a mile through a gauntlet of people with farm implements, clubs, and bamboo poles. During this wild scene, several of their blows succeeded in breaking ribs and badly damaging my right knee. The mob scene ended when they halted me in front of a Russian truck, which was used to transport me to Hanoi. During the ride they kept me face down, which allowed me to staunch the flow of blood from my face and arm. The ride itself seemed to last less than an hour. They stopped in front of an old French building of large size and allowed me to sit up in the early morning twilight. I was then unceremoniously pushed off the truck flatbed, falling about six feet to the pavement, where I suffered a shoulder separation. I was unable to move from where I had landed, and was then dragged by two soldiers into the prison yard of what I was to discover was the Hanoi Hilton.

CHAPTER 4 | ACT TWO

I was placed in a small solitary room in a section of the Hilton known as Heartbreak Hotel. Because of spinal compression and other injuries, I was unable to move for the first three days of my stay in solitary. On Christmas Day I was able to sit in an upright position and with great effort to stand for a minute or two before getting woozy. During this period I was subjected to several minor beatings and kicking sessions by my captors in efforts to force me to my feet and to persuade me to talk to them. They must have realized the futility of their efforts, for these sessions stopped just before I was able to sit and then stand. I was unable to eat a Christmas meal, but drank the pot of tea which accompanied the meal.[7]

At this point on the 21st, with the Guam force reverting to cyclic operations, Lt General Johnson informed the staffs that the raids would continue indefinitely, and that the work schedules of all units would, be structured to insure support for this extended commitment. This was easier said than done, as the General knew only too well. From a support aspect, it would mean that people on the Rock would now be fighting two wars. As General Johnson put it several months later:

You see, what bothers Maintenance is change. If you're in a cyclic operation—meaning that a given commitment is spread over a 24-hour period—you can sustain this, and that's really the thing our maintenance manning is best suited for. However, if you go into a compression and you stay in that compression, meaning that you launch and recover more or less at the same time every day, then you can stay in that compression. What causes Maintenance problems is being in a cyclic operation, and having to go into a compression, then come out of a compression back to a cyclic, and then maybe back into a compression. That's where you lose sorties; but, if you can stay in one or the other, then you can continue it. The thing that helps Maintenance (is) more or less to do the same thing every 24 hours whether it be cyclic or compressed.[8]

The staff spent the afternoon of the 21st talking to crews and asking for their ideas on how to improve tactics when and if the bombing in the North resumed from Guam. A crew which had been shot down was deploying back to the States, and their short layover on the Rock provided an exceptional debriefing opportunity. From the debriefings and crew crossfeed, with added staff suggestions, recommendations on improving current tactics were developed and forwarded to 8AF and SAC Headquarters.[9] SAC relayed some of these to the bomb wings in the CONUS for immediate application against the Eglin Test Range simulated SAM sites. A speedy analysis was vital.[10]

The Bac Mai Airfield, not far from "Killer Site VN-549," was put out of action by these strings of bombs. Targeting was so precise as to inflict substantial damage to the adjacent Bac Mai Storage Area.

There was unanimous agreement that tactics and routes should be varied so that the enemy defenders could not establish a pattern and predict routes of flight or altitudes.

Some criticism had been expressed concerning the time lag between when a suggestion was made to change a tactic or procedure and the time SAC actually approved the recommendation. This criticism does not appear to be warranted when considering the lead times required to (a) evaluate a multitude of proposals funneling in not only from the SAC bases but also from the Tactical Air Command, (b) plan a mission, and (c) coordinate support force requirements. The mission routing and the broad tactics had to be "locked in concrete" approximately 42 hours before takeoff time. Thinking back to the first three days, this meant that when the comments from the crews from Wave 1, which flew on the first day, were received from the debriefing and transmitted to SAC for review, Day Two's briefings were soon to start. Moreover, there were only approximately 34 hours remaining before Wave I on Day Three would be airborne. In addition, the analysis of the results of the inflight tests conducted by the selected CONUS units took time.[11]

CHAPTER 4 | ACT TWO

There were tactics of other types that SAC Headquarters and 8AF gave local commanders the authority to evaluate and change on the spot. One of these concerned the B-52 tail light concept developed by the 307th at U-Tapao. This involved letting the B-52D gunner use his Aldis lamp, a portable spotlight with variable filters. By monitoring his gunnery radar he knew the relative position of the aircraft behind him. At periodic intervals, especially on the bomb run, he would flash his Aldis lamp in the direction of the aircraft behind him. This gave the trailing pilot another visual reference to help keep the formation together.[12]

Some problems surfaced that were easily correctable. For example, B-52 cell designations had historically been named after colors. When launching a cell only every few hours, it was easy to set up a pattern to assure that no two sounded alike. In fact, of all the cell designators used, only a few were phonetically close. And so it happened that during the crush of assembling the first days' missions, Green and Cream Cells were inadvertently placed in the same wave on the 20th. On one other occasion, two cells had identical designators. That happened the first day. Major Bill Stocker's Rose 1 led the launches from the Rock. Later that night, in Wave III, Capt Hal Wilson and his crew were shot down while flying as Rose 1 out of U-Tapao.[13]

The crewmembers who bailed out and were recovered by Search and Rescue (SAR) forces on the first day provided valuable lessons on ways of improving rescue communications and coordination. Direct contact with the SAR detachments in Thailand also provided additional information for the crewmembers that improved their chances of being successfully picked up if they were forced to bail out over enemy territory.

The crossfeed of information from the analysts, crewmembers, planners, and Seventh Air Force (7AF) and Navy support forces was producing modifications to routing and strike tactics—two of which were already in effect for this day's strikes.[14] Release time intervals between cells were compressed. Base altitudes and altitudes between cells were changed. Also, for the first time, the cells attacking Hanoi were to fly on across the high threat zone without making the post-target turn, thereby flying "feet wet" to the Gulf of Tonkin for egress routing.[15]

These tactics and a rapid succession of even more innovative procedures applied by all the air forces, coupled with the enemy's rapidly disintegrating defensive network and resupply capability, would measurably lower the loss ratio for the rest of the campaign. We are obliged to say again that, considering the unavoidable time lag from command decision to sortie execution, the complex logistics of coordinating the actions of all strike and support aircraft, the nominal losses on Day One, and total lack of losses on Day Two, the timeliness of changes in tactics at this point in the campaign was impressive. To have made battlefield modifications any more rapidly than was actually done would have meant doing so on gut reaction or impulse. Both of these have their place in warfare, but they are usually hazardously out of place in a mass coordinated effort involving diverse air and support arms of both the Air Force and Navy. That we were able to modify and improvise our actions as quickly as we did is positive testimony to the imagination, ingenuity, and resourcefulness of the planner and aviator alike.

As the pattern of an indefinite campaign began to unfold, routings and tactics were only a portion of the subjects that came under review. Critical assessments were also made of the selection of targets.

Targeting had initially been focused for maximum psychological and logistics impact. However, the stunning number of SAM firings and the persistence of poor visual bombing conditions dictated a move more in consonance with the perception of General Meyer and the SAC staff. Something had to be done about the missiles.

General James R. Allen, then the Assistant Deputy Chief of Staff for Operations, SAC, recalls part of the review process dealing with the SAM threat:

> *One aspect of the eleven-day effort which I thought had considerable significance, involved the SAM storage sites. . . . It was on day three that we realized that, although the individual SAM sites seemed to have an inexhaustible supply of SAM missiles, current photography indicated no spare missiles at the firing sites. This implied centralized storage and distribution points and, based on the available road and bridge system in the Hanoi-Haiphong area, one could almost predict the location of the storage facilities.*
>
> *We asked SAC Intelligence to start looking for the centralized storage sites on existing photography and within about eighteen hours they started to find them. SAC immediately requested, through JCS, that these storage sites be added to the approved target list. It took another 24-36 hours to get the approval but, with one exception, all of these facilities eventually were added to the approved list. The exception was in a heavily populated area on the edge of Hanoi and the intelligence which indicated that it might be a storage facility was somewhat speculative. . . . We started hitting the storage facilities [on December 26th]. Thereafter, we continued to attack these facilities. Post strike photography of these targets showed high bombing effectiveness, and it was our strong belief that the destruction of these targets essentially had disarmed the North Vietnamese by the end of the eleven days.*[16]

CHAPTER 4 | ACT TWO

From debriefings such as this one with Capt James D. Simms and his crew from Seymour-Johnson Air Force Base, North Carolina, information was gathered to help improve tactics on future missions.

Up to this point in LINEBACKER II, everyone on the Rock had literally been going "with throttles wide open." The sheer volume of activity kept certain emotions and some of the peripheral aspects of warfare in the background. These moved to the fore as Andersen briefly transitioned back to its more customary role of ARC LIGHT support in South Vietnam.

The two things which hit home the hardest were the losses and the press coverage—and the attendant confusion surrounding both. The known losses were difficult enough to take, as they have always been, but the unknown preyed at least as heavily on the minds of the participants. No one in the crew force had the means at his disposal to conduct a head count. Even if he did, it would have been inconclusive because of Andersen aircraft forced to recover at U-Tapao or because of successful SAR efforts. Meanwhile, the press was responding to North Vietnamese claims of substantial success against the B-52s, quoting loss figures two and three times above actual. Added to that was the more personally haunting, but completely false, charge that "carpet bombing" of urban areas was slaughtering countless civilians.[17] The capstone of the latter charge was the news that the campaign was being criticized on the international scene, with especially harsh criticism from back home.[18] Men were laying their lives on the line, while performing in the finest traditions of the military, and were being made scapegoats by some for their efforts. It took maturity to face these assaults upon emotion and character. The maturity was there when needed.

95

Meanwhile, various necessary actions were bringing home the realities of the previous days' events.[19] As thoroughly researched losses were made official, the requirements for notification of home units and families came into play. Real-time world press coverage on occasion caused official notification to come after the fact, but the time-tested policy of notifying next of kin on the basis of certainty was not abandoned by the military.

Inquiry officers were appointed to research the affairs of those crews who were known to be downed, and their living quarters were locked and sealed. This silent symbol put the stamp of finality to the situation for anyone who might have had lingering doubts or hopes, and perhaps brought realization home more forcefully than any other act—short of being an eyewitness to a shoot-down.

So far, 11 B-52s had been downed: eight had gone down in the near vicinity of Hanoi and three had held up long enough to get their crews back to friendly territory. Of the eleven, six were Gs from Andersen. Of the remaining five Ds, four were from U-Tapao, with the one Andersen D being flown out of hostile territory before bail-out.[20]

DAY FIVE—WORK, COOPERATION, AND PREPARATION

Day Five, Friday, December 22d, saw a continuation of the activities of the 21st at Andersen. The commanders now faced the task of writing those letters that every commander dreads most.[21] Intermingled with that unpleasant duty were the continuing skull sessions on recommended changes that were coming from the crew force and the staff. Some of the suggestions looked good on paper, but were too complex for a formation of Stratofortresses to execute. The enthusiasm of the crews in bringing forth ideas was encouraging even when the ideas were not usable, however, and all inputs were reviewed.[22]

Another brainstorming session was held with the tactical evaluation crews and the most experienced instructor crews who had flown during the first three days. One area of discussion concerned the procedures for cells which had an aircraft abort prior to the IP. Should the cell continue as two ships, or should it catch up to the preceding cell and go across the target as a five-ship formation? Some of the pilots thought that five ships would be unmanageable when the cell was maneuvering. Others insisted that a two-ship cell would be too vulnerable to SAMs, and that five airplanes gave better mutual ECM support.[23] All reasonable ideas, along with the pros and cons of each, were again sent to 8AF for review. Eighth repeated the process using its own tactical evaluation crews and staff.

The five-ship cell issue was one of the thorniest of all, due to so many complex variables, and was late in being resolved. Several two-ship cells out of U-Tapao would later prove that this issue was more critical than expected.

CHAPTER 4 | ACT TWO

The tactics flown out of U-T on the 22d reflected the insertion of ideas from previous days into the battle plans. For the first time, the approach and egress routes were both over the water. The targets were in the Haiphong area, which represented a change in the target area emphasis.[24] The first four days were a concentration on targets surrounding and protecting Hanoi. Now the attacks were going after the Petroleum Products Storage (PPS) and the railroad support structure in the port city. The 307th SW again launched 30 sorties for these raids, as on Day Four.[25]

As the Haiphong attack materialized, it must have put the defensive command and control net into a near state of shock. Every one of the 30 aircraft was bearing in off the Gulf from the south, but the cells were fanning out on three different tracks. By the time they were approximately 60 miles south of the targets, they occupied the whole southern quadrant. At this point, the split force abruptly altered course along six different tracks, which were staggered in time and distance to provide time over target spacing. None of these tracks was aimed directly at the Haiphong complex. Three of them feinted into the coastline as though to pass well to the south of the city. Hanoi looked like a possible target for them. Then, when 30 miles or less from their intended release points, every cell again abruptly zeroed in on the city.[26]

The chaff which had already been laid by the F-4s, the preemptive Navy strikes against SAM sites, and the concentration of support aircraft undoubtedly gave a first-rate clue as to what was coming, but the sudden focus of attention by the entire bomber force initially oversaturated the system.

Only 43 SAMs were observed by the bomber force, a remarkably low number considering both the level of defenses erected around Haiphong and the fact that this was the first day of B-52 strikes against the city. However, it spoke well for the Navy's SAM suppression, which was so impressive that it prompted a personal compliment from CINCSAC.[27]

The pattern of SAM launches showed an increase in intensity and quality as the bombers pressed the attack, with the later cells experiencing the heaviest pressure and remarkable enemy accuracy. This suggested the presence of a disciplined, competent defense force, which was probably denied a B-52 hit by the innovative pretarget tactics.[28]

Although Walnut and Red Cells got "hosed down" pretty good going against the Petroleum Products Storage area, they made it through unscathed.[29] For the second day of the war, no BUFFs were downed, and it marked the first day without even any battle damage.[30] It was a turning point, and provided some overdue good news for the U-Tapao crews, who had been steadily under the gun.

97

LINEBACKER II | A VIEW FROM THE ROCK

22 DECEMBER 1972

B-52 CELLS/TARGET TIMES

'D' GUAM	'G' GUAM	'D' U-TAPAO	
NONE	NONE	SNOW	0450
		GOLD	0452
		YELLOW	0456
		EBONY	0459
		RUBY	0501
		AMBER	0503
		WALNUT	0508
		RUST	0510
		RED	0514
		IVORY	0516

LEGEND

- - - - - - - CHINESE BUFFER ZONE
△ APPROXIMATE SAM COVERAGE
TARGETS
→ BOMBER ROUTE IN
→ BOMBER ROUTE OUT
COLOR CALL SIGN OF CELL

TARGETS

1 HAIPHONG RAILROAD	12
2 HAIPHONG PPS	18
	30

65 SUPPORT AIRCRAFT

EB-66 & EA-6B (NAVY) ECM
F-4 CHAFF
F-4 CHAFF ESCORT
F-4 MIG CAP
F-4, B-52 ESCORT
F-105 IRON HAND
F-4 HUNTER/KILLER

98

CHAPTER 4 | ACT TWO

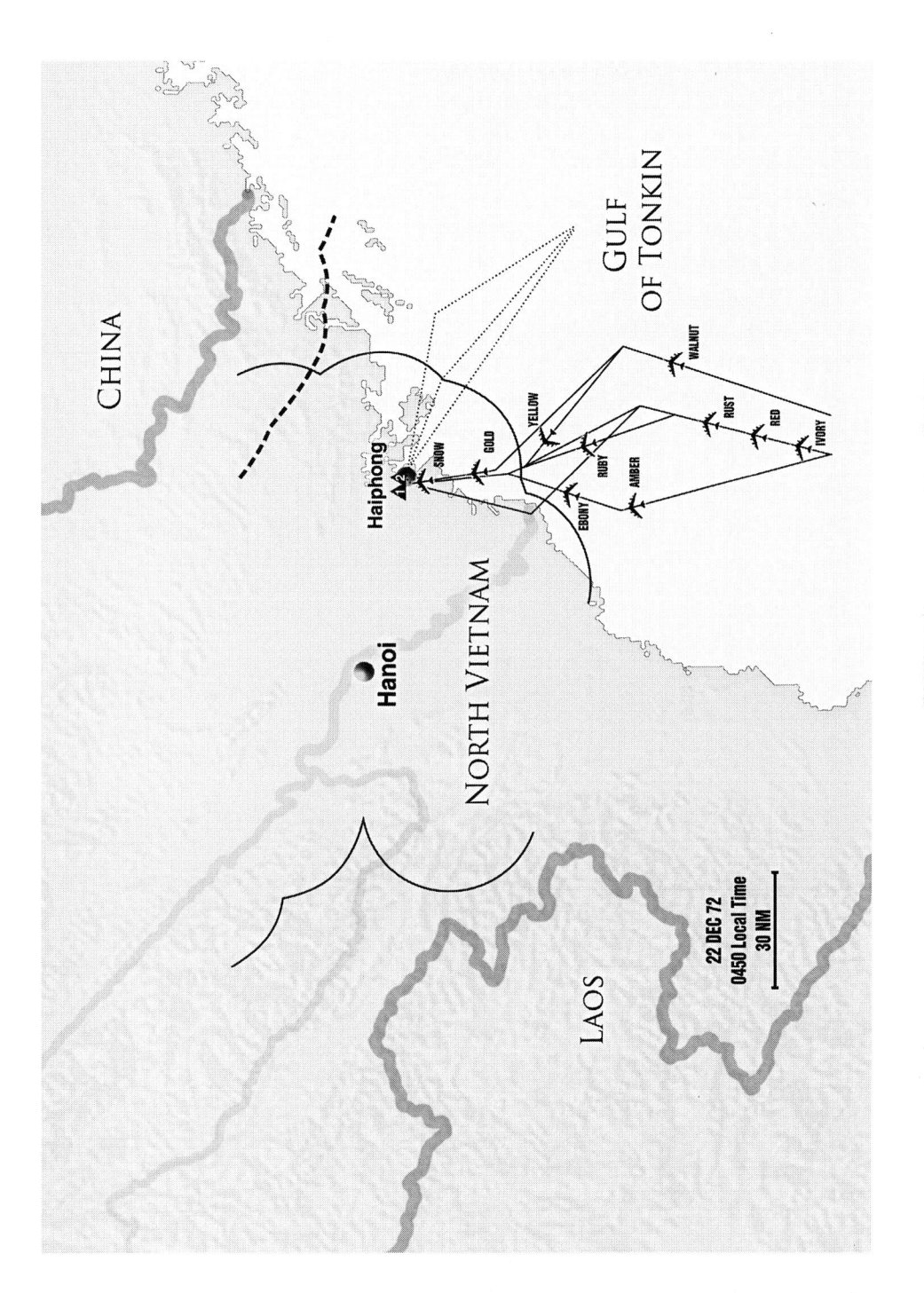

LINEBACKER II | A VIEW FROM THE ROCK

All 28 Andersen sorties, 22 G models and six Ds, again went against targets in the South.[31] This continued break on the 22d gave the command structure on the Rock a chance to do some management analysis of the effectiveness of the missions to date, and time to refine and improve operations and planning factors in the light of the indefinite nature of the campaign.

As mundane as it sounds under the circumstances, wars don't shut off the paper machine. It had become a mountain on many desks. Fortunately for the 43d SW, they had an exceptional executive officer in the person of Major Don Aldridge, who was given free rein to cut at least one mountain down to size. Actually, that was a secondary challenge, his primary charge being to insure that the target materials were on board the aircraft prior to takeoff.

Normally, a crew would have these materials to study during the general and specialized briefings, thus validating accuracy and completeness of the data. Because of the continual late arrival of the frag orders with their last-minute changes, it was physically impossible to give the crews the material until after they had reported to the aircraft.

There was one complete set of materials available for the briefing and target study by the navigation teams in the ARC LIGHT Center; the actual inflight bags would still be in the assembly process.[32]

Trailing an exhaust plume, a BUFF leaves the crowded Andersen flightline behind on its way to North Vietnam.

It was Captain Mary Speer's job to insure that the Top Secret target materials were processed and placed in the proper folders, which were then issued to the crews. It required three large black oversized briefcases to handle the material for each crew. If 30 aircraft were being launched, it would require 90 briefcases for the primary crews, plus 54 bags for the six spare crews. Each spare crew was usually required to be prepared to fill in for three different cells, hence required three times the normal target materials. Capt Speer and her intelligence specialists successfully defied the clock day after day. Sometimes they would work 36 hours without a break. Despite these long hours, they never failed to meet their extremely short suspenses.

Out on the ramp, Maj Aldridge had to supervise the delivery of every bag to as many different aircraft as the mission strength called for. To complicate his task, Charlie Tower would send crews to different aircraft if their assigned bird developed problems during preflight. There were 155 B-52s spread over five miles of unlighted ramp. A time or two it got so close that bag delivery was followed by a hasty retreat so the aircraft could take the active runway. However, there was never a late takeoff due to a delay in delivery of target materials to the crews.

A specialist concentrates on repair of electronic components in the nose of a B-52D.

Colonel McCarthy took advantage of the slackened pace on the 22d to thank the maintenance men out on the flight line for their tireless support. Some of their planes were 17 years old (22 years old at this writing—and still front line combat aircraft). The Stratofortress is as complex as it is big. It has, for example, ten independent hydraulic systems. On the B-52D these systems, as well as the four large AC generators, are powered by hot air turbines that rotate in excess of 11,000 rpm. The air for these turbines is bled off from the 16th stage compressors in the jet engine. The temperature of this air is about 250 degrees Celsius. The ducts that carry the air to the turbines are routed next to control cables, fuel tanks, and oxygen lines. A leak in one of these lines is a potential disaster for an aircraft and crew. It took considerable maintenance effort to insure that these and other complex systems operated properly.[33]

A few words of thanks go a long way with the crew chief and his assistant, or the sheet metal repairman who put the patches on the flak damage. Since crews flew different aircraft all the time, these expressions of appreciation were genuine, and reflected a spirit of close cooperation which extended from the top to the bottom of the force.

Col Thomas M. Ryan, Jr., the commander of the 303d Consolidated Aircraft Maintenance Wing (CAMW) and an experienced pilot himself, helped promote the harmonious relationship which existed between operations and maintenance. A difficult job faced him and his supervisors. Out of the 5,000 people assigned to his wing, 4,300 were TDY—living in cramped, mostly unairconditioned quarters. Away from home, they had to produce under the tight time schedules which were surging and changing. Keeping them informed was shown to be a most effective work incentive. Another incentive was some good old-fashioned fun.

Colonel Thomas M. Ryan, Jr., and the tropical Santas of the 303d Consolidated Aircraft Maintenance Wing.

Information was spread via the grapevine by having flight crews who had flown on the raids talk to maintenance work call formations. These after-the-fact sessions were supplemented by having maintenance members sit in on the predeparture briefings. This sense of involvement produced exceptional results.

CHAPTER 4 | ACT TWO

Fun took the form of an idea by Colonel Ryan to have a tropical trip by Santa Claus. The base wives' clubs provided money, along with volunteers, to buy and prepare bags of candy, nuts, and fruit. Santa's sleigh was a GI flat-bed truck, and his sack was the truck bed filled with 10,000 bags of treats. True to Santa's tradition of making dreams come true, some very pretty and shapely ladies added to the scenery by acting as his helpers. It was typical of the many minor actions generated on the Rock to keep morale high.[34]

Day Five at Andersen closed with the force gearing up for a split effort the next day. The battle was to be rejoined.

DAY SIX—BACK TO ACTION

Saturday, December 23rd, had been scheduled for six months as the Officers Wives Club formal dance. Life goes on. The women had worked hard on their floor show and, since the TDY troops could use the entertainment, the decision was made to have the dance. The night's mission, in which 12 Andersen Ds would join up with 18 aircraft out of U-Tapao, was scheduled to be airborne four hours before the festivities started.[35]

Six of the Andersen aircraft would join up on the tail end of the U-T force and attack the Lang Dang Railroad Yards 45 miles north of Haiphong. While they were doing that, the other six Andersen birds would be striking a series of three SAM sites 30 miles north of the city.[36] No one on the Rock knew why these particular SAM sites had been selected. They were not close to any of the target areas and thus far they were no threat to the force. In fact, they turned out to be among the favorite SAM sites, because the operators were poor shots. The crews speculated that those sites were the North Vietnamese version of "F Troop." They were afraid that if the sites were badly beaten up, they might be replaced by some sharp shooters. It wasn't until the night of December 26th that we were to learn the reason for this night's work.

This particular target grouping presented some interesting challenges. Because of their proximity to the Chinese border buffer zone and the SAM sites north of Haiphong, the axes of attack and withdrawal were limited to a narrow corridor coming in off the Gulf of Tonkin.[37]

The tactics for the missions against the SAM sites were unusual. Since the cells would have to fly directly over these sites to bomb them, mutual ECM protection would be marginal at best. Once an aircraft penetrated within a certain radius of a SAM site it entered a "burn-through" zone. Radar burn-through meant that at very close range the enemy ground radars were so much more powerful than the airborne ECM transmitters that they overpowered the jamming signals, permitting the aircraft to be seen on the ground radars.[38] Because burn-through would unavoidably occur, another innovation was added.

103

23 DECEMBER 1972

B-52 CELLS/TARGET TIMES

'D' GUAM		'G' GUAM	'D' U-TAPAO	
BUFF	1910	NONE	COPPER	1915
PAINT	1912		TOPAZ	1917
			COBALT	1920
HAZEL	1929		PLAID	1922
GRAPE	1932		MAPLE	1924
			SMOKE	1927

LEGEND

- - - - - - CHINESE BUFFER ZONE
▲ APPROXIMATE SAM COVERAGE
TARGETS
→ BOMBER ROUTE IN
← BOMBER ROUTE OUT
COLOR CALL SIGN OF CELL

TARGETS

1 LANG DANG	24
2 SAM VN 660	2
3 SAM VN 537	2
4 SAM VN 563	2
	30

70 SUPPORT AIRCRAFT

EB-66, EA-6B (NAVY) & EA-6A (MARINE) ECM
F-4 CHAFF
F-4 CHAFF ESCORT
F-4 MIG CAP
F-4, B-52 ESCORT
F-105 IRON HAND
F-4 HUNTER/KILLER

CHAPTER 4 | ACT TWO

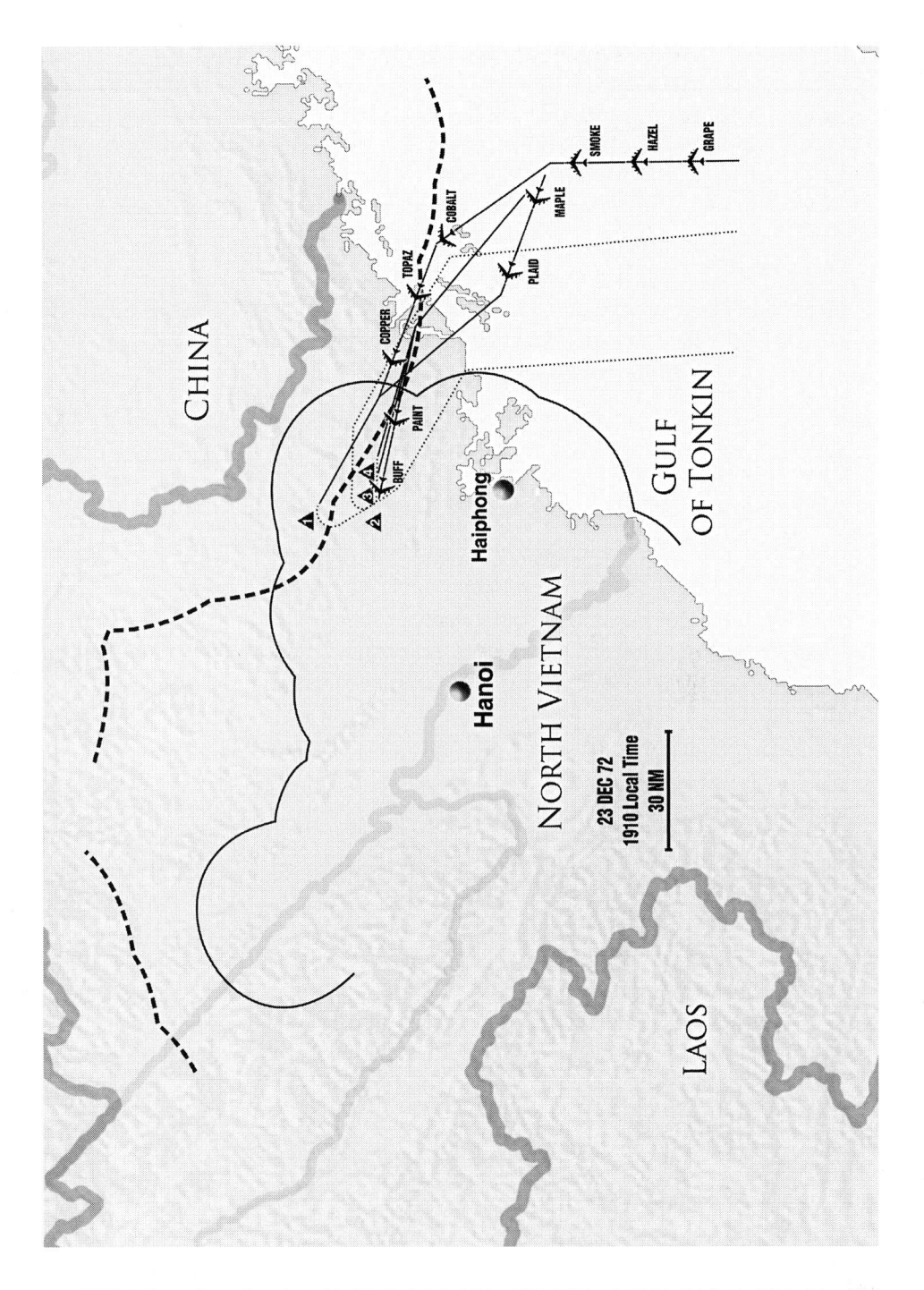

For this strike only, the cells split up into separate aircraft. One aircraft in a cell hit one site while another plane from the same cell hit another site. The second cell did the same thing, with the number one aircraft hitting the same SAM site as the number one aircraft of the first cell, and so on. By this time there was preliminary evidence that some of the NVN gunners were holding back and "going to school" on the first cells so that they could zero in on subsequent cells.[39] It was hoped that they would mistake the first single aircraft crossing the site as a cell enroute to some other target. When the bombs fell, it would be too late to realize they had been fooled.

After bombs away the individual aircraft, now split up and intermingled by cells, were to descend a substantial amount, make incremental turns, and rejoin in formation.[40] For fighter aircraft, such a tactic is no problem. But for a heavy bomber, at night, it is a difficult maneuver, especially if not well practiced.

Col Martin C. Fulcher was selected to be the ABC.[41] He was the Vice Commander of the 57th AD on Guam, and had previously flown a tour in fighters over NVN. He had already seen "his share" of SAMs before he got to the Rock.

The briefing and launch went very smoothly. Even the frag orders came in on time. This had all the earmarks of one of those missions where everything fell into place as planned.

That night the club was packed. After a week of continuous work, it was refreshing to socialize. The floor show had just started when the command radio net came alive. The wing commanders were to report to the 8AF command post immediately.

There was a serious problem, and decisions had to be made in a hurry. The problem: the support forces would not be able to rendezvous with the bombers in time to be in position for the agreed SAM suppression and MIG CAP, and the bombers were approaching the last point at which they could be recalled.

The support forces had been counted on heavily to keep the SAM sites occupied until late in the bomb run. It was impossible to calculate the odds of the bombers going in alone, spread out as they would be.

Each wing commander was asked for his recommendation. The force from the 43d which was now moving toward the recall line had been picked personally for this mission by the wing commander. They were considered among the best crews in the wing. Although the B-52Gs from Colonel Tom Rew's wing were flying in the South that night, he was also asked for his recommendations. Both commanders recommended the mission be continued. Generals Johnson and Anderson concurred, and after consultation with SAC Headquarters, transmitted the decision. As it turned out, there were no losses or damage to any aircraft on this raid.[42]

CHAPTER 4 | ACT TWO

Andersen's rejoining of the air campaign in the North was by no means the extent of the day's activities. As mentioned, the Gs were out in force in the South—36 aircraft in all, flying in the customary cyclic cell launches. Additionally, one cell of Ds went south with them.[43] So, while LINEBACKER II involvement per se was not extensive, the combination of the wave-launched 12 aircraft inserted into the cyclic cell launches caused the same sort of stresses on the support and maintenance complexes as had been seen on the 20th, on a slightly reduced scale.

Finally, due to the heavy strain on the U-Tapao crews and the losses of crews and aircraft at the Thai base, the 43d sent 22 "D" crews to U-T during the day. These additional crews helped to replace the combat losses. They also insured a sufficient crew force so that no crew would be scheduled to fly a high threat mission on consecutive nights, since by the 23rd, some of the 307th's crews had flown every night of LINEBACKER II operations.[44]

However, the stress of combat is clearly more complex than simply frequency or intensity of exposure. One of the crewmembers flying the relatively short missions from U-Tapao related that mission sequence normally kept a crew so busy from takeoff to landing that there was little time for reflective thinking. He noted that this was probably not the case in the missions from Guam. Viewed from the perspective of those who had done both, he was correct.[45]

While there were the necessary details of any flight which had to be attended to, including the essential air refueling, these did not consume a majority of the time on the Rock-originated missions. There were periods of time available to just sit and think—during which the tension built.

The situation lent itself to the term familiarly known as a "war of nerves. In the high altitude environment, pressing towards a target, there is precious little opportunity to dissipate the demands that adrenalin makes. You are the focal point of attention. You, engrossed in a target of your own, are the opposition's target—and you remain their target. You respond electronically, or by trying to outguess other electronics. It is terribly impersonal in its mechanics, but just as terribly personal in its outcome.

Then, for the fortunate, there is the flight back to base. It is the same distance as the flight inbound, approximately the same length of time, but somehow "much longer." The body, once mobilized for a fight in which it could not participate, now gradually drains itself. Most of this occurs in the few minutes following the assurance that the aircraft is beyond the threat zone. It is an emotionally exhausting experience. A certain numbness sets in, accompanied by those thoughts of hindsight which make us all so brilliant.

The opportunity for reflection was used to good advantage by many crewmembers to reconstruct, from their crew specialty perspective, their impressions of the various phases of the mission. This would become essential to successful analysis, as the normal fatigue of lengthy flight would be added to both by the post-combat letdown and the fact that the crews had been awake all day and all night. If they didn't get it down on paper or on the tape recorder while it was still fresh in their minds, much valuable information could be lost.

On the lighter side, the extended periods on the inbound and outbound flights were occasions for philosophizing, both over interphone and in writing. The latter generated some of the more unique American graffiti of that era. Kilroy would have been proud. Substantial portions of it are not suitable for repetition, as might be expected. Typical, though, of hundreds of expressions and anecdotes might be the wry humor of an unknown EW who observed that "ECM is a four-letter word,"[46] or the navigator who plaintively wrote: "Who is SAM UPLINK, and why is he doing these terrible things to me?"

Its external bomb racks empty, a B-52D returns home after a mission. Freed of the weight of up to 60,000 pounds of bombs and much of its fuel, the BUFF climbed to 43,000 feet to conserve fuel for the long trip home. At this altitude, even in the tropics, the cold air caused the aircraft to form contrails. (Photo courtesy Lt Col William F. Stocker.)

DAY SEVEN—AN ISLAND PARADISE?

On December 24th, word was passed confirming that there would be no LINEBACKER II sorties from Andersen for the 24th or 25th, but that both wings should be prepared for a maximum effort on the 26th, with drastically altered tactics. Wave leaders were to be the best crews on the island. Colonel McCarthy was to be the ABC for the entire strike force. Thirty

CHAPTER 4 | ACT TWO

ARC LIGHT sorties in South Vietnam were continued on the 24th, but all thoughts were focusing in on the day after Christmas.[47]

Detailed planning for the 26th started the afternoon of the 24th. The wing commanders and key staff were briefed on the targets and proposed tactics. Clearly, this was to be the most ambitious effort to date, and the tactics would be a challenge to even the most experienced crews. Timing tolerances to make the tactics work would be zero. In preparation training and previous missions, plus or minus three minutes was usually allowed as the tolerance on all navigation check points. This was considered the design limits of the equipment used for overwater navigation. On this mission, to get the desired effect of simultaneous times on target (TOT), three minutes was not adequate. The challenge was to develop procedures and pick the people who could make it work with a zero time tolerance.

Most of the crews took advantage of the reduced level of activity to rest up and get into the spirit of the season. Several of the squadrons organized impromptu parties at the beach and in the squadron areas. The clubs put on special holiday shows and adjusted operating hours to handle the extra work load. This and the last-minute "day before Christmas" bustle, which was evident among the permanently assigned families, spread a veneer over the Andersen scene which masked a grueling work load that was at that moment being carried by a host of mission planners and maintenance men. It further created the illusion to the untrained eye, such as that of the tourist, that the base was alive with the excitement of vast numbers of people having "fun" on Christmas Eve.

Speaking of tourists, there was perhaps no more vivid phenomenon on the island to point out the incongruity of fighting a war while living in the "lap of tropical luxury." The tourists were there in droves, mostly from Japan. Guam had become the "in" place and was particularly favored by newlyweds. Consequently, seeing other people having a good time in a completely carefree mood was disconcerting to many who were caught up in such opposite circumstances.[48]

Many devices were used to compensate for the mood people were in. They ranged all the way from personal mental disciplines to things so common as a little music. There was a song that the BUFF crews at Andersen adopted as their theme song. It was guaranteed to bring down the house any time it was played. No one who fought in the war should ever forget this most universal of Southeast Asian tunes—"Yellow River." There was nothing extra special about the tune, except that it was the last song the band played the first night of the raids. The next night, December 19th, there were no losses. Some crewmembers suggested the song was their good luck charm and should be played as the last song every night. When the band would play it, crewmembers would sing along or clap in time to the music. As LINEBACKER II progressed, the singing got louder and "Yellow River" was played more often each night. After the eleven days were over, the 43d SW adopted "Yellow River" as its theme song.

LINEBACKER II | A VIEW FROM THE ROCK

24 DECEMBER 1972

B-52 CELLS/TARGET TIMES

'D' GUAM	'G' GUAM	'D' U-TAPAO	
NONE	NONE	SNOW	1950
		EBONY	1952
		RED	1955
		WINE	1957
		AMBER	1957
		BLACK	1959
		RUBY	2002
		PURPLE	2005
		CHERRY	2007
		YELLOW	2010

LEGEND

- - - - - - CHINESE BUFFER ZONE
▲ APPROXIMATE SAM COVERAGE
TARGETS
———— BOMBER ROUTE IN
- - - - - BOMBER ROUTE OUT
COLOR CALL SIGN OF CELL

TARGETS

1 KEP RAILROAD	12	
2 THAI NGUYEN	18	
	30	

69 SUPPORT AIRCRAFT

EB-66 & EA-6B (NAVY) ECM
F-4 CHAFF
F-4 CHAFF ESCORT
F-4 MIG CAP
F-4, B-52 ESCORT
F-105 IRON HAND
F-4 HUNTER/KILLER

110

CHAPTER 4 | ACT TWO

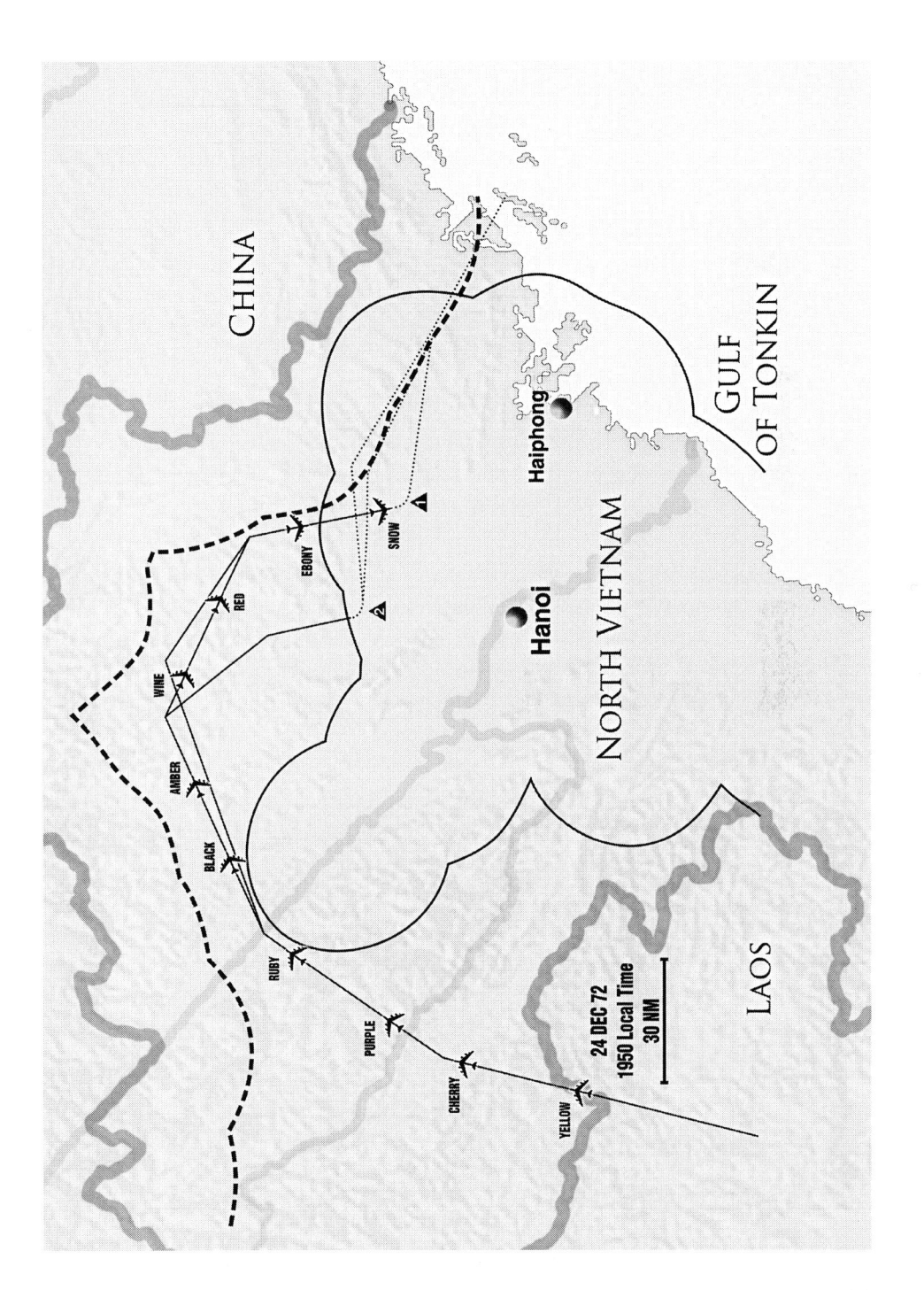

Christmas Eve at U-Tapao witnessed similar scenes. To the north in Bangkok, one of the most cosmopolitan cities in the Orient, the holiday fever was reaching its peak. U-Ts crews could visualize the bright lights and excitement there. It was only a thought. These men, having had the pressure taken off some by Andersen's sharing of the strikes of the 23d and the supplemental assistance of the crews shipped in from the Rock, slipped into their flight suits again to shoulder the entire bombardment role.

For the third straight day, targets in the immediate Hanoi area were avoided, but the strikes continued to show a diversity which kept the NVN defenses guessing.[49] After two days of penetrations from the Gulf, the bomber stream would now drive to the north across Laos to a point northwest of the NVN defenses, split in two waves, and make bomb runs on southerly headings against the Thai Nguyen and Kep Railroad Yards, 40 miles north and northeast of Hanoi. Additional tactics were used during the bomb run to confuse and deceive SA-2 operators.[50] As a final deception for the enemy gunners, the waves split in half during the post-target maneuvers, with each group exiting by varied headings and turn points. Once beyond the high threat area, all cells funneled back into a common departure route over the Gulf of Tonkin.[51]

Despite moderate defensive activity over both targets, no aircraft received SAM damage, marking three consecutive days without losses or missile damage.[52] It was not that the North Vietnamese weren't trying; rather, it appears in retrospect that the new tactics of both bomber and support forces were outpacing the defenses. The best the defenders could manage was a minor AAA hit on Purple 2, the only flak damage of the campaign. MIGs ineffectively engaged Black and Ruby Cells. Once again, it proved costly to the defenders. Another MIG went down from .50-caliber firings by A1C Albert Moore, tail gunner of Ruby 3.[53]

While it was not the first time nor last that the B-52s would fly for extended distances in-country, it is significant that the force had traversed the whole northern tier of enemy territory and emerged unharmed.

As the last sortie landed at U-Tapao in the early hours of Christmas morning, the force there would join Andersen and the other combatants in a day of peace.[54] It came on the heels of a good day's work, as subsequent bomb damage assessment photographs would prove, and was a fitting conclusion to the second phase of LINEBACKER II operations.

CHAPTER 4 | ACT TWO

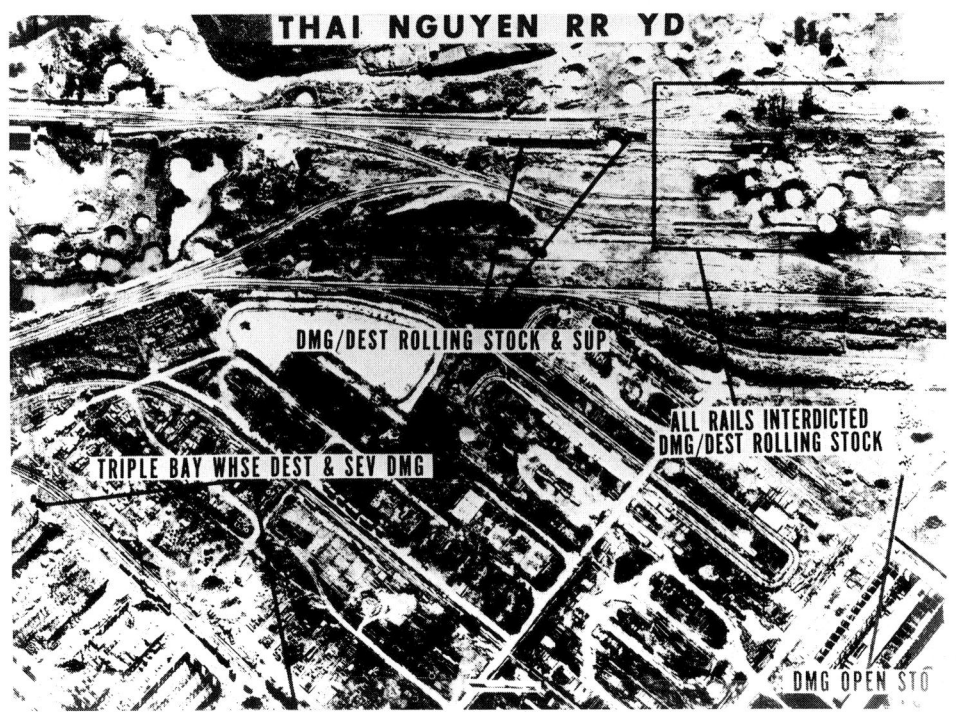

The Thai Nguyen Railroad Yard was a major transshipment point. Its interdiction was a key element in paralyzing the enemy's rail network. The inset shows some of the concentration and precision of bombardment which resulted in higher than forecast damage levels.

113

NOTES

1 *Chronology,* pp. 351-352.
2 *Ibid.,* pp. 24-25.
3 *43SW History,* p. 77.
4 *Chronology,* p. 153.
5 *Damage Analysis,* pp. A-53 to A-56.
6 *Chronology,* pp. 159-160.
7 Colonel William W. Conlee, narrative written to authors, 12 May 1977.
8 USAF Oral History Interview Program, with Lt. General Gerald W. Johnson, by Charles K. Hopkins, 8AF Historian, Andersen AFB, Guam, M. I., 3 April 1973. SECRET
9 Message (S-GDS-80), 307SW / 17AD/CC to 8AF /CC, "ARC LIGHT Compression Tactics," 22/0806Z Dec 72. SECRET. Found as Exhibit 36 to *Supplemental History on LINEBACKER II (18 29 December), 43rd Strategic Wing and Strategic Wing Provisional, 72nd,* (Volume 1), Air Division Provisional, 57th, Eighth Air Force, Andersen AFB, Guam, M.I., 30 July 1973. Hereafter cited as *43SW Supplement.* (TS)
10 *Chronology,* pp. 185-186.
11 *SAC OPS.*
12 *43SW History,* pp. 106-108.
13 *8AF History, V II,* p. 357.
14 Message (S-NOFORN-GDS-80), 7AF /SAC ADVON to 8AF et al., "Current B-52 Tactics," 21/0830Z Dec 72. Found as Exhibit 45 to *43SW Supplement.* SECRET NOFORN
15 *8AF History, V. II,* pp. 369-370. See also *Chronology,* p. 152.
16 Letter to Brig Gen J. R. McCarthy, dated 27 December 1977.
17 Senator Barry M. Goldwater, "Airpower in Southeast Asia," *Congressional Record,* Volume 119, Part 5, 93d Congress, 1st Session, 26 February 1973, pp. 5346-5347.
18 Most major national and international newspapers and periodicals, 19 December 1972 to 31 December 1972, and in lesser degree beyond that time. Two examples are "Europe Reacts to Bombing With Increasing Protests," *The New York Times,* 24 December 1972, p. 1, and "Global Outrage Mounts Against U.S. Bombing," the *Omaha World-Herald,* 29 December 1972, p. 15.
19 *43SW History,* pp. 132-133.
20 *Chronology,* p. 314.
21 *43SW History,* pp. 131-132.
22 *USAF in SEA,* p. 119.
23 *USAF AIROPS,* p. IV-288.
24 *43SW History,* p. 78.
25 *8AF History, V. II,* p. 371.
26 *Chronology,* pp. 170-172.
27 *Ibid.,* p. 185.
28 *SAC Participation,* p. K-1.
29 *Chronology,* p. 175.
30 *8AF History, V. II,* pp. 371-372.

CHAPTER 3 | ACT ONE

31 *Ibid.*, p. 37l.
32 *43SW History,* p. 40.
33 *Flight Manual,* T.O. 1B-52D-1, Air Logistics Command, Tinker AFB, OK, 1 August 1974, Excerpts.
34 *303CAMW History,* unnumbered pages of photographs between pp. 9 and 10.
35 *Chronology,* p. 190.
36 *8AF History,* V. II, p. 372.
37 *Ibid.*, p. 372.
38 *LINEBACKER II, B-52 Summary, 18-29 December 1972,* prepared by 43 SW/ DOTP, 22 January l 973, p. 1. Found as Appendix I to *43SW History.* See also *GIANT STRIDE VII,* SAC OT&E Final Report, HQ SAC/[DOXT], Offutt AFB, NE, 31 August 1971 for technical discussion of this and other SAM data. SECRET
39 *USAF AIROPS,* p. IV -272.
40 *Chronology,* pp. 187-188.
41 *Ibid.*, p. 190.
42 *8AF History,* V. II, p. 373.
43 *Ibid.*, p. 373.
44 *307SW History,* p. 81.
45 *Ibid.*, p. 54.
46 Official USAF photograph, on file at HQ SAC/OI, Offutt AFB, NE.
47 *Chronology,* p. 205.
48 *8AF History,* V. II, pp. 563, 565-567.
49 *Ibid.*, p. 374.
50 *USAF AIROPS,* pp. IV-293 to IV-294.
51 *Chronology,* p. 202.
52 *Ibid.*, pp. 206-207.
53 *307SW History,* p. 97.
54 *USAF AIROPS,* p. IV-234.

CHAPTER 5 | INTERLUDE
A MOMENT OF PEACE

For most of the crew force it was a relaxed, peaceful Christmas Day. The luster was off due to family separation, but that was to be expected. The emotional strain of the previous week's drama and hubbub had likewise taken its toll. Consequently, it was a subdued—if not somber—day for many on the Rock. Some were invited into the homes of the permanent party families; others had small get-togethers among their crew or other close friends. In the main, however, it was a time of quiet reflection.

It was a different day entirely for the people in the ARC LIGHT Center and the Bicycle Works. The pace there remained feverish.[1] Christmas was cancelled—at least for them. The moment of peace was just that—a moment. Perhaps there would be a brief word from a fellow worker, or a member of the senior staff would visit for awhile; then, on with the business. Aircraft had to be recovered, repaired, and reconfigured. Warning and frag orders poured into the Center unabated. The crush of time, which had slackened due to the previous days' modified activities, was once again upon them.[2] The mechanical, administrative, and tactical nuts and bolts of the next day's effort were being assembled.

CHAPLAINS

Christmas Day and its meaning are an appropriate backdrop for reflections on a group of men providing a form of leadership which was uniquely theirs—the chaplains. Completely overcome by the weight of numbers of those who filled the base to overflowing, the permanently assigned chaplain force was augmented by a cadre of TDY chaplains. Working in cramped quarters, and faced with a 24-hour-a-day mission of their own, they responded in the finest traditions of their calling. These men and their spiritual leadership were an inspiration to the entire Andersen population.

LINEBACKER II | A VIEW FROM THE ROCK

Loading one of 12 bombs which mounted on the external racks under each wing of the "D" model." It took teamwork, muscle power, and special equipment to position and attach each bomb.

CHAPTER 5 | INTERLUDE

Sensing their special part in the LINEBACKER II missions, the chaplains were active on the flightline. One, Lt Col Fred Lang, could be found walking from one parking revetment to another down the long line of bombers, saying a last word of encouragement or providing whatever special form of ministry that might be requested. His relationship to one of the men who subsequently became a POW was one of those little-known personal dramas. That man, touched in his own spirit and influenced by such as Chaplain Lang, left the prisons of North Vietnam to go through seminary. Captain Bob Certain, navigator on the first Stratofortress to go down in the campaign, became an Air Force chaplain himself.

CHANGE IN THE SCRIPT

As more details of the plan for the 26th were received, it was apparent that this mission was going to be the most ambitious raid to date.[3] Many of the suggestions the crews and staff had made earlier in the campaign to improve the tactics were approved by SAC Headquarters. SAC also delegated authority to 8AF to select axes of attack and withdrawal routes. General Johnson in turn asked the wing commanders to work closely with his staff to assure that wing and crew recommendations were incorporated into the overall battle plan. Eighth Air Force also delegated authority to the wings to adopt those intercell and intracell tactics they thought best for the mission. The result was a sweeping change in concept, such that it would be difficult for the uninformed to identify the upcoming battle profile and those of less than a week earlier as being tied to the same campaign. The actors and the stage were the same; the plot was changing by the hour.

The basic plan for the December 26th raid, unlike the three earlier maximum efforts, which were spread over an entire night, was to have a single mass assault of 120 aircraft striking nine different target complexes in ten separate bomber waves. All ten waves were to have the same initial time on target with all subsequent strikes completed in 15 minutes, for maximum impact on the enemy's defense network. It was hoped that this plan of attack would oversaturate his command and control system.[4] Spacings between cells, as well as altitudes for various cells, were modified in each wave.[5]

ECM tactics were changed significantly. Analyses of the ECM tests conducted in the States, plus additional reconnaissance information on enemy frequencies and techniques, gave the EWs ideas on how their equipment could be used more effectively to degrade the defenses.[6] After the analysis of the first three days' losses was completed, it was confirmed that the unmodified Gs, with their weaker ECM configuration, were much too vulnerable in the immediate Hanoi area. For the remainder of LINEBACKER II, all G models would concentrate their attacks in other areas and avoid the Hanoi complex.

Throughout Christmas Day, major changes to the targeting frag orders continued to flow into the communications center, where they were rushed to the planners. This flux required extensive replanning by 8AF and the wings. The frags eventually dictated a change to the basic plan such that the force ended up consisting of seven waves divided up against ten targets.[7] Each change had to be coordinated to determine its side effects on plans already made, and the pressure to avoid planning errors in the maze of the mass assault was enormous.[8]

The leadership and judgment of the lead crews in each wave would be critical for this mission. Careful evaluation of crew capability and performance was of great concern to squadron and wing commanders as they prepared to field a force of mixed experience levels.

Major Stocker, who had led the first B-52D wave on the first night, was again selected for Wave I lead. This wave would be exposed inside the lethal SAM lines at Hanoi for the longest time of any strike forces during the entire campaign.[9] Bucking the prevailing 100-knot headwinds, they would be easier targets for the enemy gunners—who had five days to prepare for them.

Waves II and VI, all G models, would be led by Majors Louis Falck and "Woody O'Donnell," both from Blytheville Air Force Base, Arkansas. Wave VII, also Gs, would be led by Maj Glenn Robertson from Barksdale Air Force Base, Louisiana. The total Andersen force would be 45 B-52Gs and 33 B-52Ds. The remaining 42 sorties, Waves IV and V, would come from the D fleet at U-Tapao.[10]

With force strength and crews established, the preparation continued. Seeking some semblance of the spirit of momentary peace, Colonel McCarthy invited six aircraft commanders and General Anderson to share the holiday meal with him and his wife. Bill Stocker recalled the occasion in an interview with Lou Drendel, the famous aviation artist and author of *B-52 Stratofortress in Action:*

> On Christmas night, Colonel McCarthy invited several of us to his home for dinner. We all enjoyed a real nice dinner, and we were relaxing over a drink afterwards, when the guests began to thin out. General Anderson was the first to go. He was the Division Commander, and we were still pretty busy recycling aircraft and crews, so we didn't think much of it. Then Colonel McCarthy got up and announced that he was going to have to leave also. He urged us to stay as long as we liked, and to help ourselves to his bar.... "All except you, Stocker. You're coming with me."
>
> When we got outside he broke the news to me that we would be going back to Hanoi the next day, and that I would be leading the mission and he would fly with me. He asked me to go down to the D.O.X. (the planning area), and to look over the mission profile. If I saw anything that should be changed, I should get back to him and iron it out.

CHAPTER 5 | INTERLUDE

I'll never forget the feeling of incongruity as I worked with the planners that night. I had dressed for the social occasion, and was wearing white slacks and shoes and a brightly colored shirt I had had made in Thailand. I was a sharp contrast to the uniforms around me, and the contrast highlighted the urgency of this mission.[11]

The plan of attack was for the D models from both Andersen and U-Tapao to strike Hanoi with four different bomber waves from four different directions at once. Wave I, led by Snow Cell, would come in from the northeast, hitting the Hanoi Railroad Yards at Gia Lam and the Hanoi Petroleum Products Storage Area at Gia Thuong. Wave III, Ds led by Rust Cell, would come in off the Gulf and hit SAM site VN 549 and the Van Dien Vehicle Depot from the southeast. Wave IV, D models from U-Tapao led by Pink Cell, would strike the Giap Nhi Railroad Yards from the southwest. Wave V, more Ds from U-Tapao led by Black Cell, was to hit the two targets of the Duc Noi Railroad Yards and the Kinh No Complex from the northwest.[12] Meanwhile, Wave II, G models led by Opal Cell, would be joined by one cell of Ds from U-Tapao over northern Laos, from where they would swing across and strike the Thai Nguyen Railroad Yards. Waves VI and VII, Gs led by Paint and Maple Cells, would make a double-barreled attack on the Haiphong Railroad Yards and Transformer Station from the northeast and southeast.[13]

All seven waves had the same initial TOT. The separation between targets of Waves I and V striking north of Hanoi was three miles. The same separation applied to the targets being hit by the strike force coming in from the southeast and southwest. The flight plan separation between the northern and southern bomber streams was only seven miles. This meant that 72 BUFFs would be converging on a relatively small area around Hanoi. Three miles is not much separation when one considers that Wave V, coming in from the northwest with the tailwind of 100 knots would be making good a groundspeed of over nine miles a minute. A 20-second overfly of the target would put an aircraft on a near collision course with an aircraft coming in from the northeast.[14]

To achieve the desired effect of simultaneous concentrated bombing in a relatively small area, timing was critical. The first cells in each wave were to release their bombs at exactly the same time. Navigation had to be very precise to avoid conflict with other bomber waves and to make good the timing points.

The pause was over. The brief gesture of peace had been made, both in the interests of national custom and to give the North Vietnamese time to reconsider.[15] The gesture was met with silence. However, it wouldn't be silent along the banks of the Red River for long.

LINEBACKER II | A VIEW FROM THE ROCK

26 DECEMBER 1972

LEGEND

- - - - - - CHINESE BUFFER ZONE
▲ APPROXIMATE SAM COVERAGE
TARGETS
▬▬▬ BOMBER ROUTE IN
········ BOMBER ROUTE OUT
COLOR   CALL SIGN OF CELL

TARGETS

1 THAI NGUYEN	18
2 KINH NO COMPLEX	9
3 DUC NOI RAILROAD	9
4 HANOI RAILROAD	9
5 HANOI PETROLEUM STORAGE	9
6 GIAP NHI RAILROAD	18
7 SAM VN 549	3
8 VAN DIEN VEHICLE	15
9 HAIPHONG RAILROAD	15
10 HAIPHONG TRANSFORMER	15
	120

B-52 CELLS/TARGET TIMES

'D' GUAM

SNOW	2230
SLATE	2232
CREAM	2236
LILAC	2238
PINTO	2242
COBALT	2245
RUST	2230
MAROON	2232
AMBER	2235
SILVER	2238
RED	2241

'G' GUAM

OPAL	2230
LAVENDER	2232
WINE	2235
SABLE	2238
LEMON	2241
PAINT	2230
BRICK	2233
GRAPE	2236
PURPLE	2239
COPPER	2242
MAPLE	2230
HAZEL	2233
AQUA	2236
BRONZE	2239
VIOLET	2242

'D' U-TAPAO

BLACK	2230
RUBY	2232
RAINBOW	2235
INDIGO	2237
BROWN	2240
ASH	2244
PINK	2230
WHITE	2232
IVORY	2235
YELLOW	2238
EBONY	2242
SMOKE	2245
GOLD	2245
WALNUT	2245

113 SUPPORT AIRCRAFT

EB-66, EA-3A & EA-6B (NAVY), EA-6A (MARINE) ECM
F-4 CHAFF
F-4 CHAFF ESCORT
F-4 (AF & NAVY) MIG CAP
F-4, B-52 ESCORT
F-105 & A-7 (NAVY) IRON HAND
F-4 HUNTER/KILLER

CHAPTER 5 | INTERLUDE

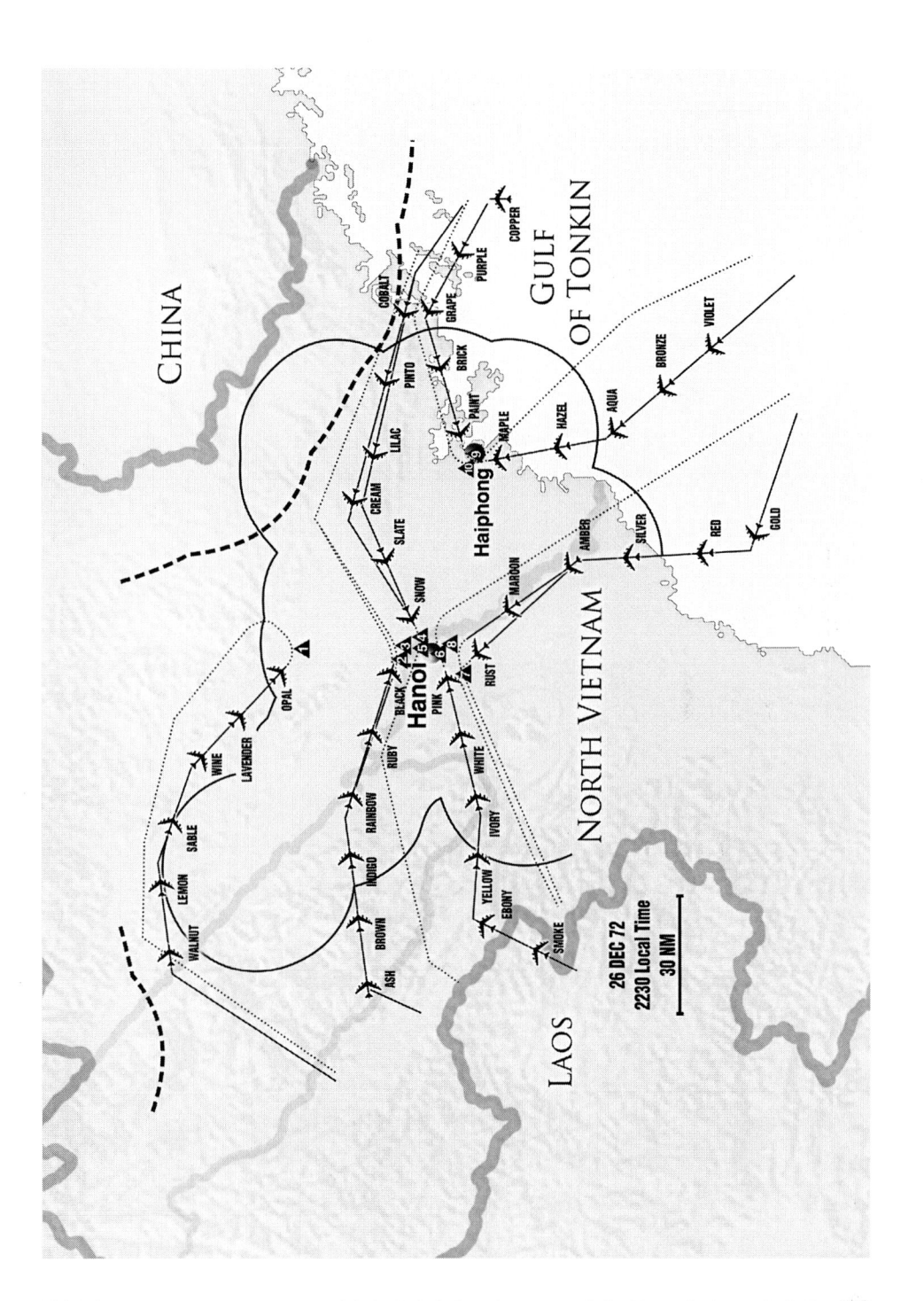

LINEBACKER II | A VIEW FROM THE ROCK

Two of the eight jet engines on a B-52 have their cowlings spread to permit maintenance specialists access to engine components.

CHAPTER 5 | INTERLUDE

NOTES

1 *8AF History,* V II, p. 376.
2 *43SW History,* p. 82.
3 *Chronology,* p. 222.
4 *USAF AIROPS,* p. IV-239.
5 *Chronology,* pp. 223-224.
6 *Ibid.,* pp. 224-225.
7 *8AF History, V, II,* p. 379.
8 *43SW History,* p. 40.
9 *Ibid.,* p. 84.
10 *Chronology,* pp. 226-227.
11 Lou Drendel, *B-52 Stratofortress in Action,* Warren, MI, Squadron/Signal Publications, Inc., 1975, p. 39. Used with permission.
12 *History of 307th Strategic Wing, October-December 1972,* Volume IV, Appendix U, Mission Charts, U-Tapao Royal Thai Navy Airfield, Thailand, 12 July 1973, Unnumbered pages. SECRET
13 *Chronology,* p. 230.
14 *Ibid.,* p. 230.
15 *Supplemental History on LINEBACKER II (18-29 December), 43rd Strategic Wing and Strategic Wing Provisional, 72nd, Volume I,* Air Division Provisional, 57th, Eighth Air Force, Andersen AFB, Guam, M.I., 30 July 1973, p. 18. TOP SECRET

CHAPTER 6 | ACT THREE
DAY EIGHT – ONE FOR THE RECORD BOOKS

Unlike the previous maximum effort missions where the launch was spread out over six to ten hours, all Andersen aircraft on the 26th would launch in one time block.[1] This meant that all crews had to be briefed at the same time, and be delivered to the aircraft at nearly the same time. The first problem was finding a place big enough to hold the briefing. The ARC LIGHT Center briefing room was big, but not that big. Even the base theatre wasn't big enough, and it was the biggest indoor facility available as far as seating capacity was concerned. The only solution was to split the briefings. Col McCarthy briefed the D model crews in the ARC LIGHT Center and Col Rew briefed the G crews in the base theatre, with staff specialists shuttling back and forth between briefings to insure continuity and coordinate last-minute information.

One moving moment in the LINEBACKER II experience occurred during the chaplain's brief portion of this pretakeoff assembly. The secret hopes of a continuation of the Christmas Day halt to the bombing had evaporated. The briefing room was once again packed to overflowing with crews, each now faced with the uncertainty of not only what this day offered, but what was in the offing in the days to come.

The chaplain who spoke, and there were several in the room because of the size of the mission, was Roman Catholic. He apologized for the circumstances and the possible affront or aggravation which his subsequent actions might create in the minds of some in the room, but he pointed out the significance of the moment. He then asked the whole body of men to please understand and to bear with him briefly. He performed a short penitential rite, and pronounced absolution and benediction to those present. Time would not permit its being done elsewhere.

An observer in the room, accustomed to the mutterings or stirrings of annoyance which are part and parcel of any large gathering, was struck by the quietness which accompanied the moment. If any there were offended, none said so publicly.

The logistics required to support this mission put unreal strains on the Combat Support Group. Anyway the schedule was handled, there weren't enough buses on the Rock to move the crews to and from the aircraft. Pickup trucks, staff cars, and anything else with wheels were pressed into service. Personal Equipment, Supply, Fuel Service, Security Police, Inflight Kitchen—these and others all found themselves with a peak workload never envisioned when the personnel and equipment package was designed during the BULLET SHOT deployment.

Down at the Bicycle Works, Colonel Ryan and the men of the 303d CAMW had their work cut out for them. The sheer size of the ramp itself presented a challenging control problem. There were over five miles of ramp pavement. When flying cyclic missions of three aircraft per launch, it was easy to concentrate the aircraft for a cell on one part of the ramp. Specialists could then be prepositioned in these localized areas to help launch the aircraft. With 78 primary aircraft plus necessary spares, this was impossible. Launch teams and specialists were spread out over the entire five miles. Just to develop a taxi plan so that spare aircraft would have access to the runway when required was a major undertaking.[2] It was to be the largest single launch of B-52 Stratofortresses on a combat mission in the history of Strategic Air Command, and it required the best efforts of all personnel.

During the pretakeoff activities, the atmosphere among the crew force was charged with emotion. The overall complexity of the substantial changes in past tactics generated mixed feelings. On the one hand, there was considerable relief and gratitude, and a surge of recovered confidence in the planners, when the news of multiple axes of attack, varied uses of altitude, and withdrawal techniques were announced. On the other hand, it was self-evident that all of these procedures, when condensed into a short time span and in a saturated airspace, would lend themselves to a major calamity if the game plan were not carefully followed.[3] The very factors which had been introduced into the script to enhance getting into and out of the target were sufficiently complicated to raise the haunting question as to the possibility of two or more aircraft inadvertently being at the same point in space at the same time. The thought was a definite incentive to pay attention to what was supposed to be done.

These were the most complex bomber tactics developed thus far during the entire war, with the most demanding navigation requirements to fly into some of the most heavily defended airspace in history. Not a single crew had the opportunity to practice those complex tactics that they would be called upon to execute flawlessly. It would take every ounce of professional ability of every crewmember to make the mission go as planned. This would be the acid test of their flying skills. However, the overwhelming feeling was one of confidence. This was the way to do it. Pull out all the stops and have at them all at once! On that optimistic note, the force launched.

General McCarthy reflects on his thoughts as he was strapping on his survival vest and parachute:

> Just before I scrambled up the crew entry hatch, I looked at the aircraft serial number painted on the side of the nose—#680. This particular aircraft had been manufactured in 1955. It had flown a lot of missions and was plainly showing its age. Seventeen years

of hard flying leave indelible marks on an aircraft that the trained eye of a crewmember can easily detect. I had seen my first D model when I was checking out in B-52s in 1960 as a junior Captain. When we went out to fly our B-52G one day there was a B-52D from another base parked next to our aircraft. The pilot happened to be there and I asked him to show my crew and me through his airplane. After we had completed inspecting the aircraft, I remember remarking to my crew, "I'm sure glad we're flying the G models and not those tired old Ds." Little did I realize then that twelve and one-half years later I would be riding one of those old D models into combat.[4]

As the aircraft taxied out to the runway, thousands on the base gathered to watch the launch. Most had had a direct part in the preparation for this moment. Many had worked 24 hours straight in preparing for this mission. They refused to go to bed. They wanted to see the results of their labors.

At 1618 local time, the first aircraft rolled down the runway. Major Bill Stocker, in his interview with Lou Drendel, captured perfectly the drama of the moment and a humorous sidelight:

Normally, we made a rolling takeoff, turning the corner from the taxiway and starting right into our takeoff run. But since I was leading and they had sterilized the runway, allowing absolutely no traffic, I requested permission to taxi into position and hold, awaiting my time out. They granted it, and in the minute or so before we took off, I was treated to the sight of one of the most awesome armadas ever assembled. As far as we could see there were B-52s lined up nose-to-tail. It's difficult to describe the feeling of leading such an array of power.

A little by-play that occurred during the launch illustrated the size of the force, and the time required to get it airborne. As I was in departure, a commercial airliner called Agana Approach Control. (Agana is the commercial airfield on Guam, and was the only other concrete available, so they were holding it open in case of emergencies.) His conversation went something like this: "Agana approach, this is Pan Am flight so and so. Request landing instructions, over." "Pan Am, this is Agana. Descend to one zero thousand, proceed south and orbit at 70 miles, 180 degree radial, Agana, and say your endurance fuel."" Agana, this is Pan Am. We have . . . ah . . . 3½ hours fuel. . . . Do you have an emergency? . . . How long a delay can we expect?" "Pan Am, this is Agana. You can expect up to a three-hour delay. No emergency. Tactical considerations!" His reaction was spontaneous, and unprintable.[5]

Two hours and twenty-nine minutes after Opal 1 started its takeoff roll, the last of the aircraft was airborne. The Russian trawler that continually maintained a position off the end of the runway also observed the launch of this strike force. His radio message would reach Hanoi long before the bombers. It was no secret that the BUFFs were coming, and there was no doubt in the enemy's mind as to what their destination would be.

LINEBACKER II | A VIEW FROM THE ROCK

Lt Col George B. Allison, radar navigator from Westover Air Force Base, Massachusetts, poses in front of a loaded external pylon on old #0100 prior to flying her on the Day Eight mission on 26 December 1972.

CHAPTER 6 | ACT THREE

Last-minute preparations for a mission continue as the crew conducts their preflight checks.

The skies over the Rock became black with the lingering exhausts of repeated takeoffs, spanning several hours. B-52G models line the ramp.

LINEBACKER II | A VIEW FROM THE ROCK

The North Vietnamese had two days to recover and reposition their forces; the main Hanoi complex had five days to recoup. They had taken advantage of the Christmas respite. Spare missiles had been brought out of storage, and new sites had been prepared.[6] Radar and guidance equipment had been recalibrated. Throughout the countryside, SAM missileers had had seven days of experience at shooting at B-52s. Hanoi, still one of the most heavily defended complexes in the world,[7] was ready and had been warned that the B-52s were on the way. Both sides knew that this night would be a test of wills.

Fourteen hundred miles away from Guam, an event was taking place that almost spelled disaster for the mission. As the tankers that were to support the B-52s were taking off out of Kadena, a C-141 inbound to Okinawa had a serious inflight emergency and had to land. This caused the runway at Kadena to be tied up for nearly 20 minutes and delayed the launch of the tankers supporting the second wave of B-52Ds (Wave III). Wave I was in the middle of refueling when the word was flashed that the tankers would be about 15 minutes late for that following wave. By the book, the mission should have been scrubbed at this point. The concept of the day's assault required all aircraft. There wasn't a 15-minute pad in the plan, and by flying the planned refueling track, there would be no way Wave III could catch up.

The ABC called Major Tom Lebar and they discussed the alternatives. If the mission was scrubbed at this point, a lot of time, money, and effort would have been wasted for nothing. If they threw the books away, violated a dozen different regulations, and moved the refueling point closer to the incoming tankers, some of the time could be made up. If Major Lebar then took his force on a modified routing with increased airspeed, while Waves I and II headed southwest past the Philippines into their compression boxes, he might possibly catch the rest of the force as it turned north. For all of this to work, Lebar's navigation team, Maj Vern Amundson and Capt Jim Strain, had to replan the course and determine the rendezvous points with the tankers and his following cells in about 15 minutes. This would normally have been a lengthy, complex navigation problem to compute and crosscheck, even on the ground. Lebar also faced the problem of compressing his wave, now spread out for refueling, without the aid of the timing boxes that the rest of the force were using. It just might work if the winds and weather stayed as predicted, all aircraft got their fuel without any problems, and if the overwater navigation was perfect.

Wave III was instructed to make the attempt. If they couldn't get in position by the time Wave I crossed the 17th parallel heading north, then the mission would have to be scrubbed. The pressure was on, and Lebar's crew knew it. Fighter and other support forces would be taking off from their Thai bases, along with their supporting tankers. The Navy and Marine planes would also be in the air prior to that time. It would be a massive air armada which would have to turn back if the bomber waves couldn't get together. A lot depended on the Wave III crews.

CHAPTER 6 | ACT THREE

Lt Colonel Allison, radar navigator in Silver Cell behind Major Lebar, recalls the incident:

We had to have the gas, and now the situation was a mess. A smooth flowing stream of bombers, coming in from the east, was scheduled to meet a like stream of tankers moving in from the north, in predetermined groupings at predetermined times over a specific point in the ocean. Only now, there was a 15-minute discrepancy in arrival times.

In a well-coordinated move, each cell within the wave began a series of airspeed and course adjustments to compensate for the problem, systematically relaying this action backwards and forwards in the stream, and passing it to the tanker force. The aircraft would now theoretically meet at the adjusted points and times, but the problem did not end there.

For one thing, the principle of effecting an inflight rendezvous contains elements of geometry and algebra, wherein the physical act is an artful blending of mathematical inputs. Except that, now, many of the entering arguments had changed. Speeds and angular relationships were no longer according to plan. They varied substantially from the norm in most cases. This meant for many crews that it all boiled down to each individual cell of aircraft, be it bomber or tanker, having to improvise their navigation to be where they said they would be, when they said they'd be there.

At first consideration, it would appear that the electronic capabilities of both types of aircraft to monitor one another's position would obviate such a problem. That, unfortunately, was not the case. For, in the confusion of the timing conflict, the natural drive for each cell to join up with its counterpart, and the close spacing of all cells, the resulting electronic rendezvous signals being emitted from the many aircraft literally saturated the scene. The rendezvous beacon signal was transmitted at a common frequency, and the result was that a radar scope might easily record twelve or more individual signals at one time. To put it in the jargon of the trade, it looked as though someone had dumped a bag of popcorn all over the scope.

However, in spite of the pressure and consternation which the timing disparity had created, all rendezvous were consistently effected and the needed fuel transferred.

Now, another problem immediately developed. A whole swarm of strike and support aircraft were supposed to be converging towards a mass effort objective, with the total B-52 force scheduled over all targets inside of a fifteen-minute time span. And there we sat, right in the middle, running fifteen minutes behind schedule. The seriousness of the pre-refueling situation was now focused on only one point of consideration—get to the target exactly on time. Otherwise, all of those pre- and post-target maneuvers and procedures could end up being as much of a hazard as the enemy's defenses. The motivation was there, and we bent the throttles as much as we dared.

Fortunately, the mission planners had continued to go to school on the experiences of the first few days, and had developed a supplemental time control box in-country to aid in refining times over target. On paper, it looked much like the cooling coils in an air conditioner, with the aircrews' responsibility being to follow as much or as little of the "maze" as necessary to arrive at the exit point of the box at a specified time prior to the target time. Along with the airspeed adjustments we had made, it worked beautifully for us, even to the accommodating of a major problem.[8]

As each wave coming in from the south crossed into South Vietnam, it started the complex maneuver to get the bombers compressed into the required formation. Although there were a few thunderstorms in the area, the compression maneuver was completed as planned. The compression maneuvers for this mission, as described by Col Allison, were not the normal tactic, and there was concern about the new crews being able to handle these without prior practice. This mission and the one scheduled for the next day would use nearly every available crew on the Rock, and insertion of new crews into the force was mandatory. For one of the new pilots, this mission was only his fourth sortie as a B-52 aircraft commander.

The common routing into the compression box, followed by a movement of a substantial portion of the force north towards the Gulf of Tonkin, created a memorable sight. The bomber crews who witnessed it will surely never forget it, observing something which may never be repeated. As one commented:

I never fully realized just how many of us there were up there, or how close together we all were, until we headed north over the Gulf. It looked like a highway at night—nothing but a stream of upper rotating beacons as far as I could see. It was sort of eerie, too, once we went into radio silence procedures. Nothing was said, but each aircraft was flashing an "I'm here" to his buddies. Then it occurred to me that we would be meeting a whole bunch more of the force which was coming up using a route over the land mass. As many of us as there were, the U-T troops were also going to be there in strength. At that moment, it dawned on me just how special this night was.[9]

General McCarthy recalls the action in Wave I:

As we headed north over the Gulf of Tonkin, I heard Tom Lebar call in that his wave was at the join-up point on time and that the wave was compressed. They had done one hell of a fine job.

When we crossed the 17th parallel, we were committed. That was the last point at which I or higher headquarters could recall the forces. From here until the target area we would be using radio silence procedures. The only radio call allowed would be if you got jumped by a MIG and you needed MIG CAP support.

CHAPTER 6 | ACT THREE

As Haiphong passed off our left wing, we could see that the Navy support forces were really working over the SAM and AAA sites. The whole area was lit up like a Christmas tree. We could hear Red Crown issuing SAM and MIG warnings to the friendly aircraft over Haiphong. We hoped that this activity would divert their attention from our G model bombers, who would soon be arriving. Even though they weren't going to downtown Hanoi anymore, they were headed for the port city. As we all knew, that was plenty tough duty.

We coasted in northeast of Haiphong and headed for our IP, where we would turn southwest toward Hanoi. The IP turned out to be in the same area that Marty Fulcher had led the BUFFs on the 23d against the SAM sites that had the reputation of being such lousy shots.

The flak started coming up when we made our first landfall. Once again, we were most vividly aware of the heavy, black, ugly explosions which characterized the 100 mm. Even at night, the black smoke from these explosions is visible. Since we were at a lower altitude than we had flown before, our wave would be more vulnerable to this AAA than on most previous missions. Close to the IP the flak became more intense and the explosions were closer to the aircraft.

As we turned over the IP we picked up the first SAM signals. We could see them lift off, but their guidance seemed erratic. The SAMs exploded far above us and at a considerable distance from the formation. It appeared that F Troop was still in business and their aim was as bad as it had always been.

However, inbound to the target the SAM signals became stronger. Capt Don Redmon, the EW, reported three very strong signals tracking the aircraft. Bill Stocker ordered the cell to start their SAM threat maneuver. The navigator, Maj Bill Francis, reported that we had picked up the predicted 100 knot headwinds.

Then the SAMs really started coming. It was apparent this was no F Troop doing the aiming. The missiles lifted off and headed for the aircraft. As we had long ago learned to do, we fixed our attention on those which maintained their same relative position even as we maneuvered. All of the first six missiles fired appeared to maintain their same relative position in the windshield. Then A1C Ken Schell reported from the tail that he had three more SAMs at six o'clock heading for us. The next few minutes were going to be interesting.

Now that the whole force was committed and we were on the bomb run, I had nothing to do until after bombs away, so I decided to count the SAMs launched against us. Out the copilot's window, 1/Lt Ron Thomas reported four more coming up on the right side and two at his one o'clock position. Bill reported three more on the left side as the first six started exploding. Some were close—too close for comfort.

Listening to the navigation team on interphone downstairs, you would have thought they were making a practice bomb run back in the States. The checklist was unhurried. Capt Joe Gangwish, the RN, calmly discussed the identification of the aiming point that they were using for this bomb run with his teammate, Major Francis.

About 100 seconds prior to bombs away, the cockpit lit up like it was daylight. The light came from the rocket exhaust of a SAM that had come up right under the nose. The EW had reported an extremely strong signal, and he was right. It's hard to judge miss distance at night, but that one looked like it missed us by less than 50 feet. The proximity fuse should have detonated the warhead, but it didn't. Somebody upstairs was looking after us that night.

After 26 SAMs, I quit counting. They were coming up too fast to count. It appeared in the cockpit as if they were now barraging SAMs in order to make the lead element of the wave turn from its intended course.

Just . . . prior to bombs away, the formation stopped maneuvering to provide the required gyro stabilization to the bombing computers. Regardless of how close the SAMs appeared, the bomber had to remain straight and level.

One crew during the raids actually saw a SAM that was going to hit them when they were only seconds away from bomb release. The copilot calmly announced the impending impact to the crew over interphone. The aircraft dropped its bombs on target and was hit moments later. That's what I call "guts football."

At bombs away, it looked like we were right in the middle of a fireworks factory that was in the process of blowing up. The radio was completely saturated with SAM calls and MIG warnings. As the bomb doors closed, several SAMs exploded nearby. Others could be seen arcing over and starting a descent, then detonating. If the proximity fuse didn't find a target, SA-2s were set to self-destruct at the end of a predetermined time interval.

Our computer's bombs away signal went to the bomb bay right on the time hack. Despite the SAMs and the 100 knot headwinds, the nav team had dropped their bombs on target at the exact second called for in the frag order.

Some minutes afterwards, as we were departing the immediate Hanoi area, there was a brilliant explosion off to our left rear that lit up the whole sky for miles around. A B-52D (Ebony 2) had been hit and had exploded in mid-air. Momentarily, the radios went silent. Everyone was listening for the emergency beepers that are automatically activated when a parachute opens. We could make out two, or possibly three, different beepers going off. Miraculously, four of the Kincheloe Air Force Base, Michigan crew escaped the aircraft,

CHAPTER 6 | ACT THREE

becoming POWs. Then there was a call from another aircraft, Ash 1, stating that he had been hit and was heading for the water. The pilot reported that he was losing altitude and he was having difficulty controlling the aircraft. Red Crown started vectoring F-4s to escort the crippled bomber to safety.

As we withdrew farther from the target area, the gunner reported an additional barrage of SAMs headed our way. Bill gave the order to the formation to again start their maneuvers. It seemed like an eternity before the gunner reported that they had gone over the top of the aircraft and had exploded. That was our last encounter with SAMs that night.

Now came an equally hard part—sweating out the time until the entire bomber stream had dropped their bombs and the cell leaders reported their losses. From the congestion on the radios it was apparent that the NVN had loaded up plenty of missiles and were using them.

Suddenly, one of the cells in our wave reported MIGs closing in and requested fighter support. Red Crown, who had been working with Ash 1, started vectoring other F-4s to the BUFF under possible attack. I gave the command for all upper rotating beacons and all tail lights to be turned off. As the F-4s approached, the MIG apparently broke off his attack, because the fighters couldn't locate him and the target disappeared from the gunners' radars. This appeared to be another one of those cases where the MIGs were pacing the B-52s for the SAM gunners. It was speculated that if, while doing this, they thought they saw a chance for a one-pass quick kill, they would try to sneak within range and fire off a missile. Either that or make one screaming pass through the formation and then disappear. It was apparent that they didn't want to mix it up with our F-4 escort.

Finally, the last cell had exited the threat zones and reported in. The customary expression of this was, "So and so cell, out with three." A more picturesque expression, which sort of captured what was happening, was when a formation reported themselves "over the fence with three". Except for the violent loss of Ebony 2 and the problems Ash 1 was still having, the rest of the force was intact. Considering what had just happened, their successive reports of "out with three" were heartlifting. Cream 1 and 2 had the dubious honor of both being damaged by the same SA-2 detonation, but it was minor in both cases and they were headed for the Rock.

As we turned south, we could overhear Ash 1's conversations. He had made it to the water OK and was now heading south. Red Crown was giving him the position of friendly ships in the area. However, Capt Jim Turner reported the aircraft seemed to be flyable and he was going to try to make it to U-Tapao. There were probably a couple thousand guys who were listening that were praying he would make it. He almost did. He crashed just beyond the runway at U-T, a tragic loss after so heroic an effort. Only the gunner and copilot survived the crash, and the copilot would not have made it without the bravery of Capt Brent Diefenbach, who had himself landed only a few minutes earlier. His quick thinking and ingenuity enabled him to reach the crash site, where he pulled 1/Lt Bob Hymel from the wreckage. TSgt Spencer Grippin escaped the burning wreck when the tail section broke free on impact.

After I had given my ABC summary report to 8AF over HF radio, I noticed I was having difficulty breathing. Although we were supposed to still be at combat pressurization at this point, I couldn't stand the pain any longer. Before takeoff, the flight surgeon had put a stethoscope to my lungs and said the pneumonia, which he had diagnosed a week earlier, had settled in both lungs. He went away shaking his head, muttering something about idiots and pilots who fly with pneumonia.

With normal cabin pressurization, 100 percent oxygen, and pressurized oxygen flow, the pain subsided to tolerable levels. It would take me six months to get over the effects of the pneumonia.

As we headed towards Okinawa and our post-strike refueling, made necessary by the extraordinary length of this mission, we received a piece of bad news from our tankers. The weather had deteriorated in the refueling area. Visibility was dropping and there was moderate turbulence at refueling altitude.

By this time it had been about 16 hours since the crews had reported for duty. They had just flown one of the most difficult and dangerous missions of their careers, and now they would have to conduct a night refueling in weather conditions which were very hazardous. If we had planned it, we couldn't have come up with a situation that would more severely challenge the ultimate flying abilities of these combat crews. However, despite visibility that sometimes blotted out the view of the tanker's wings and the refueling lights, and turbulence that made the instrument panel hard to read, all refueling was completed as briefed.

As the lead airplane touched down safely at Guam in the warm sunshine, I felt proud and humbled to have been their commander. The record mission was executed flawlessly, and the crews met every challenge thrown at them. When the history of Air Power in Southeast Asia is finally written, the raid flown on 26 December 1972 by the B-52s and their support forces will, I suspect, be judged as one of the most successful bombing missions of the war. The credit for this outstanding achievement belongs not only to the magnificent flight crews, but to all the support and maintenance troops as well. It was truly an outstanding team effort. [10]

CHAPTER 6 | ACT THREE

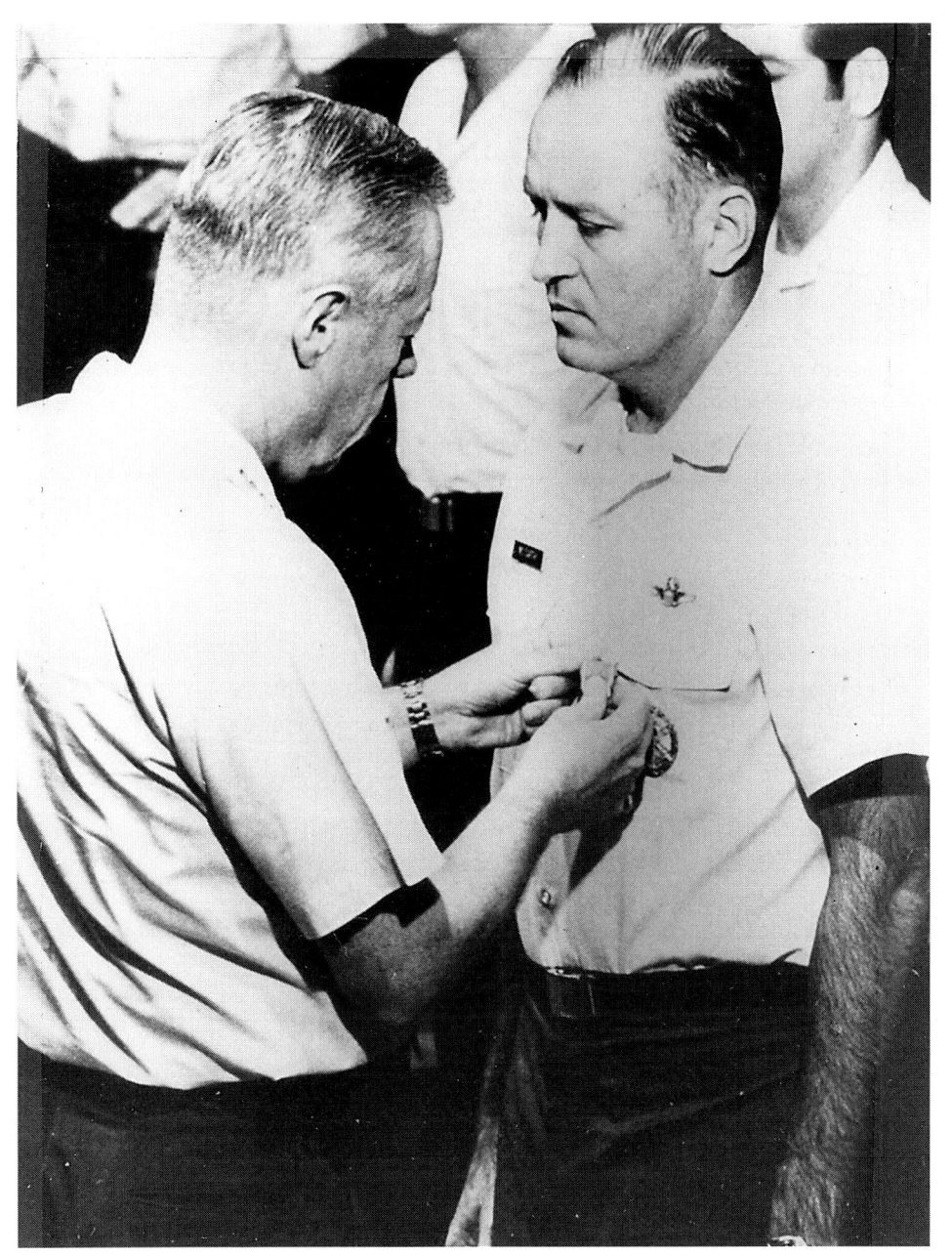

Gen John C. Meyer, Commander-in-Chief, Strategic Air Command, presents the Air Force Cross to Col James R. McCarthy, Commander, 43d Strategic Wing, for his participation in the historic mission of 26 December 1972.

TOTAL FORCE PARTICIPATION

No day better lends substance to General McCarthy's closing sentence than does December 26th and the effort it generated from the crews and aircraft committed to direct support of the B-52 strikes. It represented, with a record of its own, a perfect example of what happened every day.[11]

Throughout the campaign, each day's support figures varied, based on weight of B-52 effort, target area, axes of attack and withdrawal, threat analysis, and so forth. Generally, the battle plan called for a fairly consistent ratio of support to bomber aircraft. The overall ratio, support to bombers, averaged out to 1.3 to 1. Consequently, this day of maximum compressed effort produced an unprecedented concentration of special-purpose sorties over a narrow span of time.

One hundred and thirteen support sorties were distributed over North Vietnam and the Gulf.[12]

Type	Function
F-105	Iron Hand
A-7E	Iron Hand (USN)
F-4	Hunter/Killer
F-4	Chaff
F-4	Chaff Escort
F-4	MIG CAP
F-4	MIG CAP (USN)
F-4	B-52 Escort
EB-66	ECM Support
EA-6B	ECM Support (USN)
EA-6A	ECM Support (USMC)
EA-3A	ECM Support (USN)

The air discipline required of the bomber crews in precisely executing maneuvers and controlling their times and positions in the saturated airspace was mirrored by the skill of the support combat crews. Probably never before had such a large airborne jet arsenal of diverse talents and specialties been orchestrated to perform as a body. This uniqueness was imbedded in the bomber crews' minds, and the evidence is convincing that it was similarly regarded by the support crews as well. Hence "superior performance" was the identifier for virtually every combatant that day.

The authors, as personal monitors of the radio nets on Day Eight, state with conviction that this same identifier extended outward from the target zones to encompass those ground, sea, and airborne personnel whose monitoring and coordinating of the effort was not only essential but also masterfully accomplished.

We speak throughout this monograph of air refueling support for LINEBACKER II aircraft. No day more emphatically illustrates the wholehearted dedication of the KC-135 force than does this same December 26th. Due to the around-the-clock nature of refueling support operations, tying them to a calendar or clock day becomes a problem, but the 26th provides a good approximation, and a definite indicator of exceptional effort.

From Kadena, 95 sorties provided 156 refuelings, setting an all-time record. More refueling sorties were to be added. SAC tankers from U-Tapao, Takhli, and Clark Air Base, Philippine Islands, recorded 607 air refuelings solely for TACAIR aircraft, the largest number provided during the eleven days.[13] This total of 763 refueling sorties to support one mission was the largest such undertaking in the war.

These statistics, and interpretations of parallel statistics, must be viewed in the context of one invariable. Only 194 KC-135 airframes were in the theater at the time, and some of these were not available due to required maintenance. It is evident, anyway the statistics are worded, that it took a Herculean tanker effort to support the extensive December 26-27 operation. The maintenance people delivered the KC-135s and the crews delivered the gas. No one appreciated that accomplishment more than the fatigued crews winging their way back to the Rock.

Meanwhile, back in Hanoi, Lt Col Bill Conlee was recovering from the ordeal of his first few days in prison:

> On the evening of December 26th, the walls of the Hilton shook with the proximity of the B-52 bombing, which was a great encouragement to me. It was immediately obvious that SAMs were being fired from several sites located right outside the prison walls. I could also hear AAA fire coming from the roof of the Hilton, using the prison as a sanctuary from attack. The next morning a large group of ashen-faced Vietnamese came into my room and asked how close the bombs were and what airplanes were dropping them. I said, "Very close, and you know already," and smiled. This proved to be a mistake as I was quickly subjected to a rough beating, the worst I received during captivity. From this experience I concluded that the North Vietnamese were genuinely terrified about the B-52 bombing and were striking out in fear and frustration at an available target—me. After licking my wounds for a couple more days, I was then moved to a larger room in what was known to POWs as Heartbreak Annex.
>
> In the new room, I was reunited for the first time with other POWs, all recent B-52D and B-52G shoot-downs. The seven POWs with whom I was now quartered included Lt Col Yuill and Lt Mayall from our crew. From them, I was relieved to hear that our entire crew had reached the ground safely and all were okay except for shrapnel wounds and one

bamboo knife wound. I also discovered that all the senior officers in this room had spent approximately one week in solitary, which seemed to be the modus operandi at this time.

Our new situation afforded several of us an opportunity to clean up some for the first time since shoot-down, as the Vietnamese brought all the dishes from this portion of the camp to our room for washing. This was an obvious attempt to humiliate us and to make us lose face. However, we took advantage of this situation to gain intelligence as to how many recent shoot-downs were in our area, and how many were sick and could not eat even the skimpy meals of turnip soup which we were being served. We were also able to peek out through the door of this room and observe the bomb shelters in the Hilton courtyard and to see other POWs from time to time. We were subjected to daily lies from our captors about vast numbers of B-52s shot down, to which we responded with laughs. Even the Vietnamese were hard pressed to keep a straight face while spinning many of their propaganda yarns. This daily ritual also included morning and night "camp radio" broadcasts, from a squawk box on the wall, which were nothing but propaganda speeches.[14]

Old #0100 drops another load of bombs on the enemy. A very reliable aircraft, she survived LINEBACKER II without a scratch. After the campaign was over, a crack was discovered in the wing which was beyond economical repair. She was decommissioned and put on display at Andersen Air Force Base as an ARC LIGHT Memorial dedicated to all the brave men who lost their lives in B-52 operations in Southeast Asia.

CHAPTER 6 | ACT THREE

An example of the effects of concentrated B-52 bombing. Here a military vehicle and tank truck conversion facility, located in the southern part of Hanoi, is put out of action.

LINEBACKER II | A VIEW FROM THE ROCK

Hanoi's Gia Lam Railroad Yards, like all other major transshipment points in North Vietnam, were a prime target.

Bombs impacted across the warehouses and the marshalling yard at Giap Nhi, scattering boxcars in their path.

CHAPTER 6 | ACT THREE

This low altitude photo shows the degree of damage which was characteristic in many North Vietnamese storage facilities.

DAY NINE—LAST MOMENT OF PAIN

From the debriefings of the crews on the 26th, more lessons were learned that were applied on the 27th. For example, two-ship cells weren't "hacking it" over a target defended with the intensity encountered at Hanoi. Both Ash 1 and Ebony 2 had proven that the night before. Even though they were both D models, each was a member of a two-ship cell, the third aircraft in each case having aborted enroute to the target at a point where spare replacement was not possible.[15]

From this point on, if an airplane dropped out of a formation enroute to the target, then the remaining two would marry up with the cell ahead or behind to form a five-ship cell. The use of the tail light for maintaining cell position, a proven concept, was refined and improved.

The minimum post-target turns and the expanded altitude separation between waves and cells within waves seemed to be especially effective tactics. So was the selective deployment of chaff, as numerous crews observed SA-2 detonations when the missiles penetrated the chaff clouds.[16] Many of these and other improved tactics were the direct result of suggestions made by the combat crews who had flown earlier missions.[17]

LINEBACKER II | A VIEW FROM THE ROCK

North Vietnamese antiaircraft artillery being fired from the city streets. Only one instance of minor damage was inflicted on the B-52 force by AAA.

The 27th was a replay of the 26th, except on a smaller scale and minus Haiphong. The port city had seen the last of the BUFFs. This reflected one of the problems the planners were facing: they were running out of suitable targets. The damage already inflicted on most targets was higher than the planning factors. This meant that the crews were bombing more accurately than predicted with the exceptionally precise data provided by the staff. This combination of staff expertise and crew ability produced results of remarkable quality. Also, intelligence sources noted that the North Vietnamese were not able to repair the damage as fast as had been expected.

The morale of the crew force was especially high on the 27th, following the epic experience of the 26th. Also, Andersen had finally received some bomb damage assessment photographs. The crewmembers, for the first time, got a look at the results of their work. What had not been evident through the undercasts and the darkness of the nights before was now recorded on drone, SR-71, or U-2 photography. Crews could match pictures with targets they had been fragged against. Additionally, since their cross-hair positions were evaluated on the film taken by a camera attached to their radar sets, many could work with the photo interpreters to pick out their own train of bomb craters on the ground. For these men, the sense of detachment which goes with high altitude nighttime bombardment was wiped out with one point of the finger and the simple statement, "Right on target!"

Morale zoomed farther when prayerful hopes were fulfilled, as some crewmembers shot down and initially reported as MIA were announced as being recovered. Even the newsphotos of those who were POWs were cause for rejoicing. Missing friends were still alive. Some

CHAPTER 6 | ACT THREE

excellent escape and evasion suggestions had also come from the downed crewmembers to improve chances for recovery in case of bailout over enemy territory.

The strike force for the 27th consisted of 60 bombers, 21 Gs and 39 Ds—30 of which came from U-Tapao. Andersen also flew 30 additional sorties—six in South Vietnam, plus 24 in other parts of North Vietnam. The northern strike force flew six waves, hitting seven targets, again using simultaneous initial times over target (TOT). The entire force was to drop its bombs in ten minutes instead of the 15 minutes planned for the night before.[18]

A new tactic used on the 27th was to split the wave coming in from the northeast into three smaller streams attacking separate targets, then reform into one wave after the PTT. This tactic called for precise navigation on the join-up to insure that each cell returned to its proper place in the wave after bomb release. Some cells in various positions in the wave were fragged at the same altitudes, and it was critical that they "fell back into line" at the proper place. The wave coming in from the southeast used identical pre- and post-strike tactics, except that it split into two streams from the IP inbound.

General Meyer, CINCSAC, wanted to insure that the SAM sites were destroyed as quickly as possible, even if it meant using Stratofortresses to do it.[19] He was still feeling the pressure associated with the losses of the big bombers, and was being pressed into what was, to him, a violation of basic air doctrine. One of the "first commandments" for the employment of strategic air power is to destroy enemy defenses initially; then military and industrial targets can be concentrated on with little loss to the attackers. Several missile sites, in reality pinpoint targets, had to be dealt with by the bombers. One particular site southwest of Hanoi had been especially troublesome.[20] Dubbed "Killer Site VN-549," it was a target this night. However, it survived the attack, and fired only moments after being bombed to add Ash 2 to its list of kills.

By now, most of the Andersen crews had flown at least two missions, and some had flown three. This by no means matched the U-Tapao rate of five to six sorties per crew, but was about the maximum possible, considering the sortie rate and the absence of LINEBACKER II assignments for Guam aircraft for three of the first eight days. They had joined that large fraternity of men who had flown over Vietnam's infamous Red River Valley and survived. They were battle-tested pros, and it could be seen in their actions. The atmosphere at the mission briefings was more relaxed. There was more of the good-natured small talk between crewmembers which had been seen on less hazardous missions.

Even the frag orders started coming in on time. The crews had sufficient time to study target materials during briefings, rather than at the aircraft. Base support and maintenance had also developed procedures to smooth out peak workloads. If that seemed like a long time in coming, recall that this undertaking was a scant week and one-half old.

LINEBACKER II | A VIEW FROM THE ROCK

27 DECEMBER 1972

B-52 CELLS/TARGET TIMES

'D' GUAM

GREEN	2300
COBALT	2303
TOPAZ	2305

'G' GUAM

BEIGE	2300
CHERRY	2302
CHROME	2304
CHESTNUT	2306
OPAL	2308
GRAY	2310
CINNAMON	2312

'D' U-TAPAO

RUBY	2259
WINE	2303
AMBER	2305
BLACK	2309
LEMON	2300
ASH	2300
PAINT	2302
RAINBOW	2306
SILVER	2309
IVORY	2309

LEGEND

- - - - - - CHINESE BUFFER ZONE
▲ APPROXIMATE SAM COVERAGE
TARGETS
BOMBER ROUTE IN
BOMBER ROUTE OUT
COLOR   CALL SIGN OF CELL

TARGETS

1 LANG DANG	21
2 SAM VN 234	3
3 DUC NOI	9
4 TRUNG QUANG RAILROAD	12
5 SAM VN 243	3
6 VAN DIEN SUPPLY	6
7 SAM VN 549	3
	60*

* 2017 Edition Note: The number of aircraft per target does not add up to 60. This error was present in the original draft of the book, and we have not tried to correct it.

101 SUPPORT AIRCRAFT

EB-66 & EA-6B (NAVY) ECM
F-4 CHAFF
F-4 CHAFF ESCORT
F-4 (AF & NAVY) MIG CAP
F-4, B-52 ESCORT
F-105 & A-7 (NAVY) IRON HAND
F-4 HUNTER/KILLER

CHAPTER 6 | ACT THREE

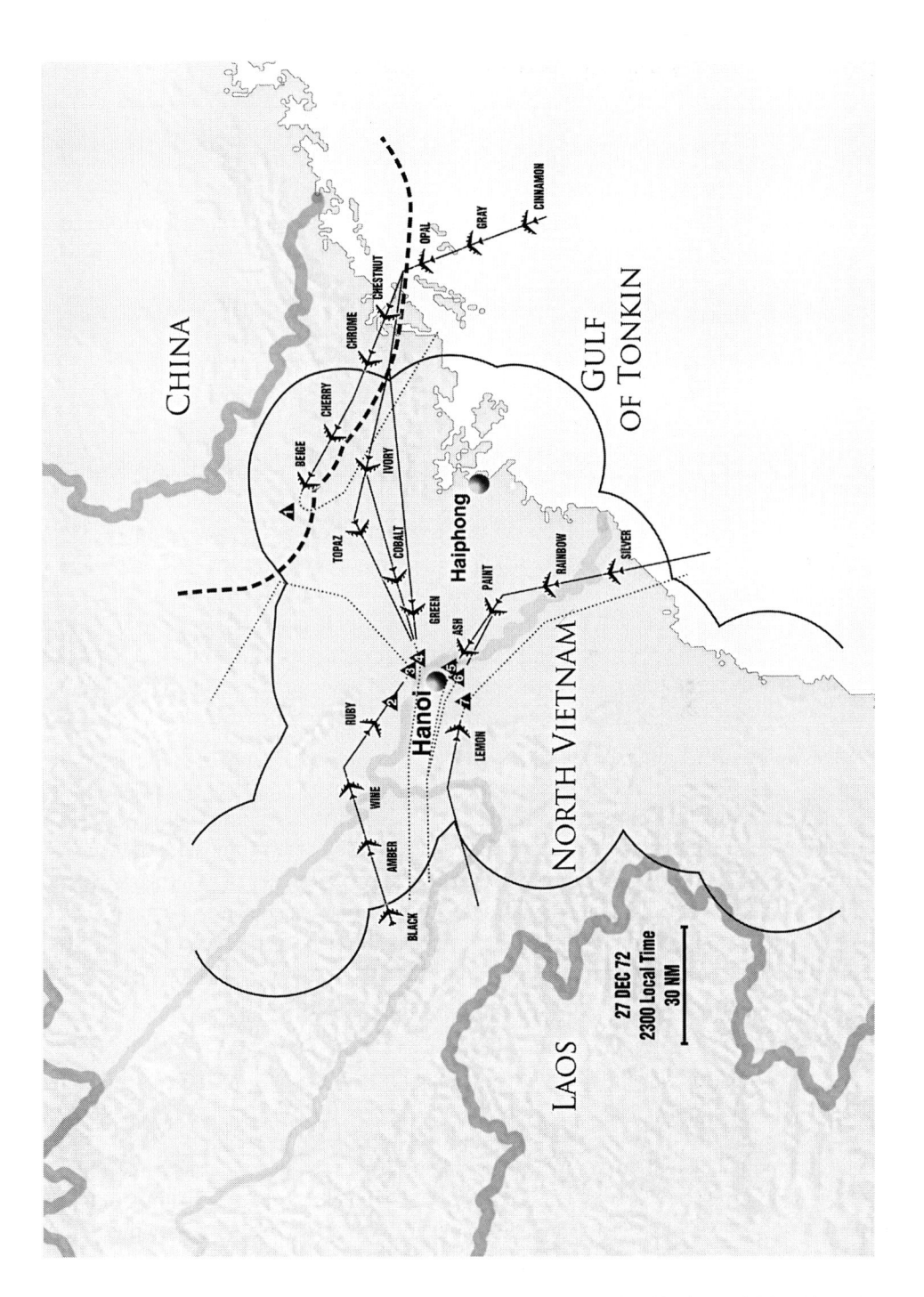

The "troops" in the Bicycle Works had outdone themselves day after day in providing high quality aircraft for the flight crews. There were fewer abort conditions and inflight equipment failures than at any time during the previous six months of BULLET SHOT activity. That pattern would continue right up to the end of LINEBACKER II, and even be reflected in performances which followed the campaign. The 303d CAMW would subsequently be selected as the winner of the 1972 Daedalian Weapon System Maintenance Award. In receiving this most prestigious of all Air Force maintenance awards, the wing was cited for superior performance in providing safe and highly reliable aircraft to support the mission in Southeast Asia.[21]

The launch on Day Nine went like clockwork. The weather was good in the refueling area and the support aircraft were blistering the defenses, or were in position to escort the BUFFs. The mission was flown as briefed.

When contacted in September 1977, Lt Col Phil Blaufuss had vivid memories of the campaign:

I flew in LINEBACKER II with one of the most fantastic pilots I've ever known—Capt Dick Martin. In fact, I had the good luck to be teamed up with five tremendous people. I would have flown anywhere—absolutely anywhere—with those guys. Our home station was Barksdale AFB, Louisiana, but we were doing most of our living as a crew "family" on the Rock in those days. Our ability to work smoothly as a crew was what made the difference between having a good tour or a so-so one, and it was a key to our success on December 27th.

That was my third mission as an RN during the campaign; in many ways, it was the toughest. That might sound surprising, considering that we had less thrown at us in the way of defenses that night. However, my job was to bomb. Worrying about defenses was, from my downstairs position, an evil I had to ignore. I can't think of anything more useless than to worry about missiles when you're an RN or NAV, stuck in the belly of an airplane with no windows to see all the hell that's breaking loose, no guns to shoot, no ECM equipment to jam with, and no control column or throttles to maneuver the plane. Talk about a waste of time; worrying about enemy defenses is sure one of them under those conditions.

The toughness of this mission came from the target we were fragged against. We were leading Opal Cell in the middle of a wave of Gs, all headed for the Lang Dang Railroad Yards. The marshalling yards and rolling stock were important targets, but Lang Dang is up in the hill country northeast of Hanoi. That's in the boondocks, and our radar aiming points were some of the toughest I ever had to use. If the mission was going to be a success, I simply had to put my full attention to my own special job. That was—to make sure the checklist for releasing armed bombs was completed with no omissions, to make sure the equipment was working properly to solve the bombing problem, and to make sure I was on the aiming

CHAPTER 6 | ACT THREE

point. Anything else would have detracted from that, and that's where Dick Martin showed one of the many talents he had. Dick kept the crew informed. No theatrics—just good, solid information. He reacted to a SAM launch with a quiet, running advisory on where it was and what it meant to us. He was relaxed and self-disciplined, and it rubbed off on the whole crew. Each one knew his job, kept the rest of the crew advised—if they needed to know—and we all calmly went about our business. I counted it a pleasure to fly with them.

We did have a situation that night which wasn't in the frag order, and caused us some aggravation. After we had coasted in off the Gulf, we temporarily became a four-ship cell. At least that's what the guys in Gray Cell behind us observed. It seems that a MIG had decided to come up and play tag-along by flying loose formation off our wing. It was probably another case of being up there as a traffic cop for the SAM batteries. Anyhow, somebody called for MIG CAP, and when the F-4s headed our way I'm told that our fourth aircraft left the scene about as urgently as anybody had ever seen before.[22]

The Lang Dang Railroad Yards were an especially challenging target for the crews late in the campaign. Located in rugged hill country only 20 miles from the Chinese border, the target and surrounding area offered very limited radar returns for the radar navigators to aim at.

A MIG tried to engage Ivory Cell, but was unsuccessful. Since Ivory's route was not that far from Opal's, and because Ivory coasted in off the Gulf first, it is possible that the pass was made by the same MIG which later joined up with Opal. There just weren't all that many MIGs out and about—that night or any other time. In fact, the MIGs were one of the more pleasant surprises of the whole campaign.[23] The TAC fighter troops and the Navy attack aircraft kept the airfields pretty well under thumb from start to finish. With very few exceptions, the crews were more concerned with keeping the formation together than with "sweating out" MIGs. The value of that one fact alone cannot ever be measured, since an integral formation proved to be such an essential element in a successful B-52 assault.

With the mission of the 27th, the handwriting was on the wall. After the period of preparation which the North Vietnamese, and especially Hanoi, had available to them to prepare for the post-Christmas attacks, and after the barrage firings of SAMs which were generally pretty accurate on the 26th, the accuracy of the 27th's missiles was noticeably poorer. More were launched than on the day before, but that number would never even be approached again.[24] The last desperate attempt to defend Hanoi was being made. As far as two crews were concerned, that attempt was successful. But it would not be repeated again during LINEBACKER II. Capt John Mize and his crew from Ellsworth Air Force Base, South Dakota, flew a D model out of U-Tapao as Ash 2. Their target was SAM site VN-243, just across town from Killer 549. According to eyewitnesses in Paint Cell, just behind Ash, Ash Cell's bombs destroyed at least one SA-2 in the lift-off stage and sent several others on completely erratic flight paths. However, their egress routing took them past 549's deadly shooters. That was all it took, and a resultant detonation wounded every member of the crew. Despite his wounds and severe damage to his aircraft, Captain Mize flew his BUFF for 48 more minutes into Laos, where he finally had to bail out his crew and then himself. All were later recovered.[25] He was awarded the Air Force Cross for his actions, and each member of his crew was awarded the Distinguished Flying Cross.

Shortly after Ash 2 was hit, Cobalt 1 sustained a nearly direct hit from another site when only seconds away from their release on the Trung Quang Railroad Yards. The EW, Maj Allen Johnson, had only time enough to call out "They've got us!" before the violent detonation near the forward wheel well and right wing root. Maj Jim Condon, a substitute RN from March Air Force Base, California, tried his best to release the weapons. When these attempts failed, Capt Frank Lewis ordered the crew to bail out. Except for the mortally wounded navigator and the EW, who became MIA, the Mather Air Force Base crew was captured and ended up in the Hanoi Hilton.[26]

CHAPTER 6 | ACT THREE

Elsewhere in the battle, Major Don Aldridge, as the Deputy ABC, was rapidly recording events and impressions aboard Green 1:

Started to see AAA around the IP or slightly before. Most of it was around 20-25 thousand feet, but fairly heavy. MIG activity west of Hanoi, but were being engaged by F-4s. TAC kept them busy and off of us. Started seeing SAM firings around the IP. Lots of low AAA, some 122mm unguided rockets. SAMs becoming more frequent and apparently aimed at the cell. F-4 escort flying either side of cell in opposite direction—rotating beacons on.

TRUNG QUANG RR YD
20 DST BOXCARS
NO BLDGS REMAIN STANDING
ALL RAILS INTERDICTED
7 DMG BOXCARS

The Trung Quang Railroad Yard was clobbered. All buildings in the target area were destroyed, and all rail lines severed.

TTR maneuver was commenced . . . Gunner reports 2 and 3 tucked in and cell moving as one aircraft. SAMs getting pretty thick but most missing by at least one-half mile. Gunner reported one SAM low very close to Green 3. Three okay. 120 seconds to go—six SAMs obviously aimed at Green cell—the cell continues to maneuver. The two SAMs at one o'clock are well off the target—exploded high about one-half mile horizontal range; the two at 10 11 o'clock are not going to be a factor. Pilot reports that the two at 9 o'clock are staying in one spot on the side window and appear to be tracking Green 1. TTR continued but altitude variation increased. SAMs are now visible to me, out of my seat so I can see. Looked close to me! At what appears to the uninitiated to be very close, the pilot starts a small rapid descent—SAMs go over the aircraft—I can see the exhaust—they both explode above the aircraft. Height unknown but the concussion can be felt slightly and the copilot reports he could hear the explosion. Back on altitude immediately and TTR continued. Actually, the pilot never ceased the "book" TTR bank angle for the entire time except for the final moments of the bomb run . . . then straight and level, on bomb run heading.

Bombs away, back into the TTR. "Ash 2 is hit!" Ash 1 reports he is heading for the fence, Ash 2 has two engines shut down and a few moments later is down to four engines and losing altitude fast—down to 19,000 and still in North Vietnam. At 1608Z (11:08PM) there was a gigantic explosion on the ground. The entire landscape was illuminated through the thin undercast. Ash 1 reports that it was not caused by Ash 2 who now has five engines operating and is across the fence but still losing altitude.

My notes say 31 SAMs fired at Green Cell up to this point—I was writing down data on damaged aircraft and probably missed several, but none of them were exceptionally close. Cobalt 2 reports he thinks he is now a flight of two—lead has taken a hit! Confirms Cobalt 1 is going down—beepers heard—in target area. Red Crown notified. Apparently hit immediately at or after the BRL.

Cinnamon Cell reports three aircraft out and that accounts for the B-52G wave within radio range. Green, Topaz, and Ivory report out with three. Gunner reports SAM low at 4:30 position—time 1620Z. Suspect it was a flare or rocket—no known or suspected site in this area. Ash 2 is nearing time to punch out. 1647Z—looks like the aircraft crashed about 20-25 miles from NKP. Can hear Rescue talking to pilot and radar navigator but cannot hear their reply. Later confirmed—all out and okay.

Return trip to Andersen—long. Copilot (Capt Tom Brown, Jr.) executed perfect penetration, approach, and landing. Flying time—15.5 hours. A thoroughly professional crew; AC (Capt Glenn Schaumberg) and RN (Capt Arthur Matson) are superb. These two among the best I have ever seen.[27]

Ash 2 and Cobalt 1 went down, but the last price had been paid. These two losses, and two incidents of minor damage that night, marked the end of NVN's defensive successes against B-52s during the campaign.[28] It was now a matter of time.

CHAPTER 6 | ACT THREE

The extreme velocities of SAM fragments (8000 fps) and the pressure differential from inside this tip tank resulted in gaping outward explosions of the sheet metal.

DAY TEN—THE END IS IN SIGHT

Debriefings from the crews who flew on the 27th indicated that some of the formations were still spreading out too much. For the 28th, the intracell spacing was decreased. Anticipating a need for even better nighttime formation flying in the target zones, selected instructor pilot crews flying the deadhead routes to and from the raids had, for several days prior, been directed to experiment with different ideas on how to safely fly a closer, more compact formation. One idea that proved feasible in the D model was to decrease the interval between aircraft until the pilot could see the glow given off by the exhaust gases from the engine tailpipes of the aircraft ahead of him. This glow appeared as eight round lights which could be used as an artificial horizon. The aircraft that was following kept his wings parallel to the tailpipes on the aircraft in front of him, thereby assuring a coordinated turn pattern and roll-out. At the same time, the trailing pilot could monitor the spacing and intensity of the glowing lights as a rough guide to maintain adequate separation. This tactic was used very effectively on the 28th.

As this new tactic was being added, another new one was being removed, reflecting a constant update and improvement in force employment. The procedure of varying the hold time from bomb release until the start of the PTT was deleted. It seemed a good concept for throwing the enemy gunners off guard, but the procedure actually caused a loss of cell integrity, putting aircraft out of position. Since cell integrity had the highest priority, the tactic was abandoned.[29]

The LINEBACKER II attacks for Day Ten called for 60 B-52s—15 each Ds and Gs from Andersen, and 30 Ds from U-Tapao. They formed into six waves attacking five targets. Four waves and their four targets were in the immediate Hanoi area, while the other two waves attacked the Lang Dang Railroad Yards. Those yards, a key choke point in the supply routes from China, got heavier attention during the last days of the campaign than any other target.[30]

Of the four targets around Hanoi, three were again SAM sites. SAM attrition rates had never reached the desired levels, due mainly to the persistence of poor tactical bombing weather over Hanoi. Throughout the whole course of LINEBACKER II, there were only 12 hours of good daylight bombing weather in 12 days.[31] Since the SAM sites were still basically intact, the BUFFs had to go after them on a continuing basis. The sites targeted were specifically tied to the ingress bomber tracks. Further, post-target maneuvers and routes were tailored to site locations and to the designated sites to be struck. As a gesture of genuine respect, VN-549's immediate area was carefully avoided.

CHAPTER 6 | ACT THREE

For this strike, even heavier chaff was sown over Hanoi; none was provided for Lang Dang. Limited and inaccurate firings had been the rule in the up-country target area, and mutual ECM support was considered sufficient.[32]

Unlike the bomber tracks on previous missions, those on the 28th crossed each other on egress from the various targets, some waves making sharp breakaway turns and some executing flyovers. This called for precise flying, since the four Hanoi targets once again had the same initial TOT. Twenty-seven aircraft bombing in the northwest quadrant of Hanoi either crossed directly over one another or flew within five miles or less of other cells on reciprocal tracks. Altitude separations had been planned to allow for these procedures, but no one in a higher wave could afford to be off altitude after bombs away without endangering the crossing bombers in the lower wave. Any high wave bomber with battle damage that couldn't maintain altitude would simply have to break clear of the bomber stream.

Another innovation this night was that the two waves attacking Lang Dang flew on absolutely reciprocal tracks, with the second wave egressing southeast directly down the inbound track of the first wave, while the first wave executed a large turn back towards the Gulf.[33] In this case, the last TOT of the first wave was ten minutes before the first TOT of the following wave.

In evaluating the evolution of tactics which occurred during the campaign, General McCarthy observed:

Although many of the proposed tactics looked relatively simple on paper, they were in actuality very complex when you remember that the pilots were manhandling 400,000 pound airplanes around the sky. Unlike a high-performance fighter or the newer B-52G and H models, the B-52D does not have powered controls. It takes a lot of old-fashioned muscle power to fly precision formation or maneuvers with that D model. Some of the D crews at that time were G and H model crews who had only the experience gained by the two-week "D-difference" course. Performance characteristics were so pronounced that some H model pilots might take as much as two months of flying the D before they felt "comfortable" in the aircraft. Flying the B-52D has been compared to driving an 18-wheel truck without power steering, air brakes, or automatic transmission in downtown Washington during the rush hour.

LINEBACKER II | A VIEW FROM THE ROCK

28 DECEMBER 1972

B-52 CELLS/TARGET TIMES

'D' GUAM

PLAID	2215
SABLE	2218
BRASS	2223
GOLD	2219
INDIGO	2223

'G' GUAM

SNOW	2215
BROWN	2217
LILAC	2219
BRONZE	2221
VIOLET	2239

'D' U-TAPAO

PINTO	2215
HAZEL	2215
PEACH	2219
YELLOW	2223
WHITE	2215
RED	2218
RUST	2223
SMOKE	2231
ORANGE	2233
QUILT	2236

LEGEND

- - - - - - - CHINESE BUFFER ZONE
▲ APPROXIMATE SAM COVERAGE
 TARGETS
 BOMBER ROUTE IN
✈ BOMBER ROUTE OUT
COLOR CALL SIGN OF CELL

TARGETS

1 LANG DANG RAILROAD	24
2 SAM SUPPORT FAC 58	18
3 DUC NOI	12
4 SAM VN 266	3
5 SAM VN 158	3
	60

99 SUPPORT AIRCRAFT

EB-66 & EA-6B (NAVY) ECM
F-4 CHAFF
F-4 CHAFF ESCORT
F-4 (AF & NAVY) MIG CAP
F-4, B-52 ESCORT
F-105 & A-7E (NAVY) IRON HAND
F-4 HUNTER/KILLER

CHAPTER 6 | ACT THREE

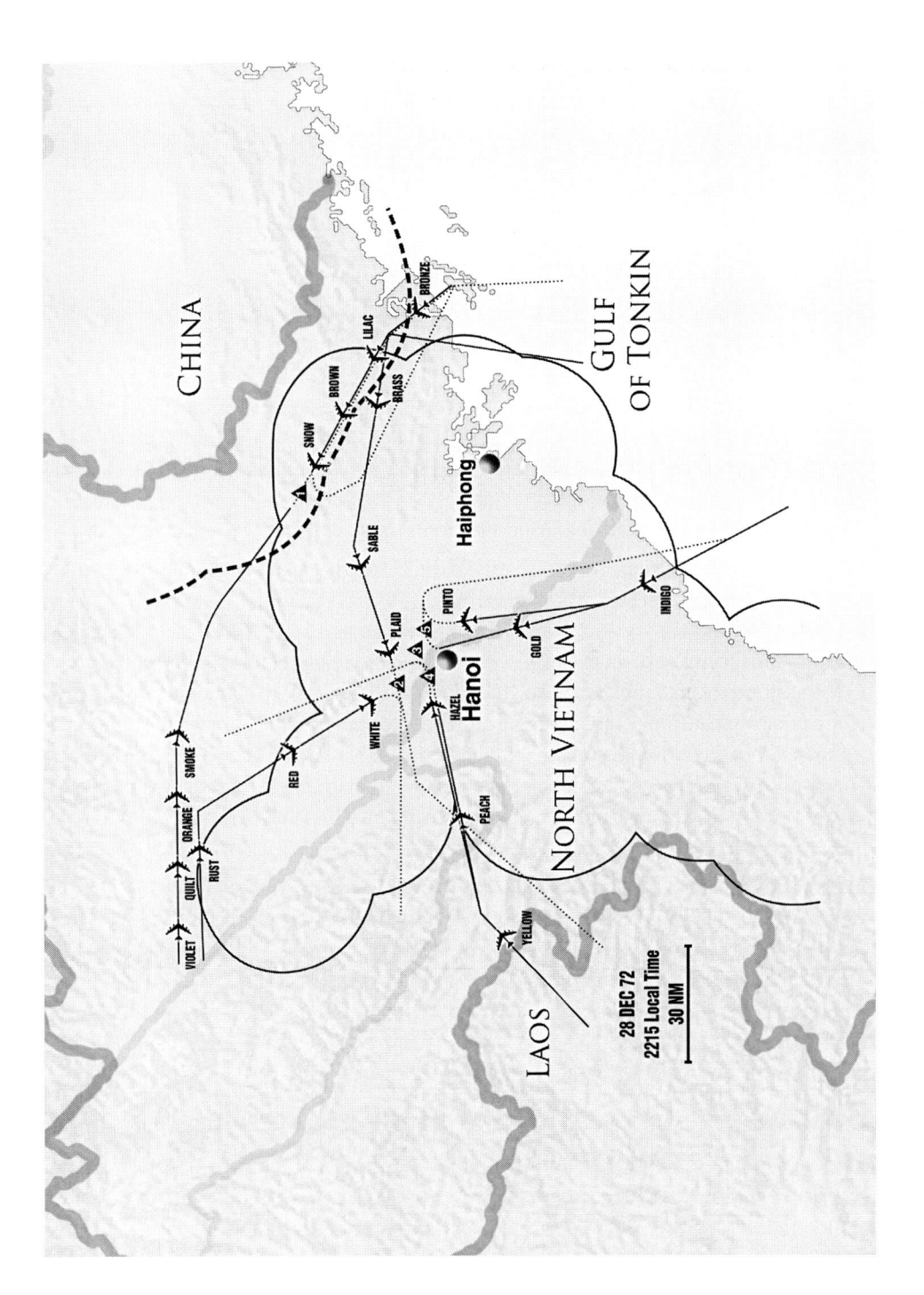

LINEBACKER II | A VIEW FROM THE ROCK

There was some real-time criticism expressed, with spill-overs of criticism continuing to the present, that SAC used stereotyped thinking and was slow to change the tactics used in LINEBACKER II. These criticisms were based mostly on limited insight, and I think that the opposite is true. In the space of eleven days, we completely revolutionized many modern day bomber tactics. Some of the complex tactics used during the latter days of LINEBACKER II would not have worked during the earlier days because the crews at that point did not have the experience to execute them. Although we had a small number of highly experienced crews at the start of LINEBACKER II who had been directly exposed to North Vietnam's defenses, the majority of our crews had never seen a SAM or AAA, or dealt with MIGs. Many of our aircraft commanders had less than 1500 total flying hours, and some had less than 100 hours' aircraft commander experience. It was a tribute to the professionalism and "can do" attitude of these relatively inexperienced combat crews that they could execute the new or revised complex tactics flawlessly the first time without ever having practiced them. In my judgment, had these tactics been tried earlier, before the crews became accustomed to the heavy SAM environment and gained additional experience in intracell formation flying, the results would have been sheer disaster.[34]

The selection of routes, the new tactics, the chaff laying, the efforts of 99 support aircraft—all worked together to achieve the desired results. Every target was successfully attacked without any losses. The crews reported that there were less SAMs fired than on previous missions, and that many of them had very erratic guidance. Most of the missiles with a steady flight profile exploded harmlessly away from the BUFFs. There were reports that some of the fighter pilots were turning on their landing lights, apparently trolling for MIGs. The bomber crews didn't see any MIGs that night, although one was shot down by MIG CAP.[35] Even the AAA was lighter and far off the mark. It was apparent that the enemy command and control network was breaking down and that they were running out of bullets.

Andersen capped off the day's efforts by sending 28 aircraft to attack targets in southern North Vietnam, Laos, Cambodia, and northern South Vietnam. They had sent 30 sorties on similar ARC LIGHT missions the day before. There was a threefold message here. First, North Vietnam wasn't experiencing anywhere near the full punishment which it might. Secondly, target damage levels had been met to the point that truly strategic targets were becoming scarce. Finally, it was clear proof that the bomber force could meet a dual commitment of theater support/interdiction bombardment and northern strategic bombardment, and do it for an indefinite period. However, the latter was not to be.

BOMB LOADERS

As the Day Ten flyers were returning to the Rock, the last of the aircraft for Day Eleven were being uploaded. Without knowing it, the weaponeers were writing their last chapter in the LINEBACKER II story. Before the last bomb was hung, they would have loaded approximately 58,000 of them, representing 18,000 tons of destruction. Over 15,000 tons of those were felt by the NVN heartland; the rest went against other theater targets.[36] These figures do not take into account the thousands of weapons pre-positioned on spare aircraft or made ready for loading in the munitions staging area. Nor do they include the uploads for Day Twelve, a day that LINEBACKER II would never see.

Loading bombs has been a back-breaking, exhausting business ever since a way was devised to carry them aloft. It is particularly so in the tropics. Scenes at Guam and Thailand, which had become commonplace over the years of the B-52 effort, were scarcely distinguishable from those of previous wars. Some of the specific machinery had changed with time, but a conventional bomb is ultimately dealt with on an individual basis, just as it has always been. The sweating, straining, and profanity associated with the peculiar cussedness of over 750 pounds of metal and explosives, apt not to do what you want it to do—at just the wrong time in a loading sequence, is an experience shared by thousands of weapons handlers through the years. Nothing had changed in December 1972. It was more of the same.

A B-52D makes its approach for landing under the watchful eyes of a security policeman.

Only now it had been intensified. What had for a long time been 12-hour shifts, six days a week—difficult enough in their own rights—had become a seven-day-a-week job. It had gone on around the clock for a week and a half at this point, with no indication of stopping.

Weapons preparers and loaders, many augmenting the force from other career fields, worked continually to insure the required supply of ordnance for the strike force.

The assembly line aspect of preparing bombs for installation on or in aircraft is impressive to watch, especially because of the methodical, rhythmic actions of the weaponeers. From the time the crates of unprepared bomb bodies arrive at the assembly area, through the point of being configured with fuses and tail components, to being positioned on bomb racks, the loading area is one constant bee-hive of activity.

Observing the repetitive efficiency of so many young men systematically going through the ritual of bomb preparation gave the observer a peculiar feeling. Common sense dictated that the repetition alone must be boring, unchallenging, or unfulfilling. Probably it was also frustrating, because the repetition was being performed in the face of a timetable which must be met.

Twelve hour shifts, seven days a week, manhandling 750-pound bombs required a lot of sweat and stamina.

But, of course, the ritual could not be allowed to become routine. A weapon suddenly not under complete control is a real "attention-getter." Much depended on the loaders to mold a seemingly unending supply of items which, after being hauled nearly 3,000 miles, were expected to work.

CHAPTER 6 | ACT THREE

For the LINEBACKER II missions, there was not a drastic change in the total volume of work, because the previous sortie rate had been high. However, it did mean a more compressed timetable for having a complete force configured all at about the same time, and it meant a smooth insertion of each special activity into the total preparation of the aircraft. Everything else came to a standstill while each specialist group did its job at the aircraft, and other people were always waiting for their turn to get at the plane and do their own work. Always the pressure. But it was worth it. The men knew where their aircraft and bombs were going. It meant something. "I'm actually putting them where they count!"

Still, these men, like all of their counterparts who stayed behind while the missions were flown, were removed from the full reality of what had been happening. Somehow, it wasn't satisfying enough just to know. They wanted to be told.

Requests were made for any interested crewmembers who had participated in the raids to go to the weapons assembly area and talk to the munitions teams. By Day Ten, several such trips had been made. Some crewmembers counted it as one of the most worthwhile off-duty experiences they ever had. They were asked to wear flight suits and were told that what the loaders wanted to hear was, very frankly, war stories, but with special emphasis on successful releases of "their" bombs. The stories weren't hard to tell.

A young munitions technician carefully attaches the arming wire of a 750-pound M-117 to the shackle mechanism.

When the crewmembers got to the munitions area, the munitions people were all gathered into a large hangar-like building. For each session, a huge group of them would gather in a semicircle, most of them sitting on the floor, while the crewmembers "held court" in the middle. After a short briefing on the missions, the targets, and some personal experiences, there were the predictable questions. "What did it feel like?" "Were you scared?" "What does a SAM look like?" "How many did they shoot at you, and how close did they come?" "Were there any MIGs?" "Could you see the bombs hit?" "Were the explosions big?" Then came questions that were more subtle, but just as filled with meaning. "Did all your bombs release?" "Was there any problem with the release system?" "Did you have any 'hangers'?" "Which tail number did you fly?" The answer to that one caused the questioners to exchange glances. You could tell that they were trying to remember if that was the aircraft that they had loaded. It was genuinely important to them.

DAY ELEVEN—THE CURTAIN COMES DOWN

The last day of LINEBACKER II was Friday, December 29th, the 11th day of bombing. In the morning, all planning and support activities were still geared to an indefinite campaign, so preliminaries for Day Twelve were already well under way. Only later in the day was General Johnson notified that all bombing operations north of 20 degrees latitude would be terminated following the mission of the 29th.[37]

Complete bomb damage assessment photography was still not available, due to seasonal weather problems. Interpretations of photography available on 12 of the targets were made to extrapolate damage levels on all points that had been struck. Desired damage levels had been achieved, the enemy was either unwilling or unable to repair the damage,[38] and there were no more "worthwhile" targets left in the immediate Hanoi or Haiphong areas.

There were, however, two lucrative SAM storage areas, as well as the Lang Dang Railroad Yards. The tactic employed against the latter was nearly a carbon copy of the double-wave strike and withdrawal that had been performed the day before. It was at the Phuc Yen SAM Storage Area northwest of Hanoi that the most complex tactic to date was flown. It provided a fitting way to climax the campaign.[39]

Three waves of three cells each attacked Phuc Yen. The release time of each cell exactly matched those of its counterpart cells in the other two waves. This meant that bombs from three separate sources were in the air at the same time, all destined for the same target. Added to that was the post-target routing, which had two of the waves crossing tracks, separated only by altitude. A delayed release by a high sortie would result in the bomb train falling through the path of the other wave. This required closer timing tolerance than was established on any previous day.

CHAPTER 6 | ACT THREE

Moreover, the wave coming in from the northwest, instead of flying a crossing track after release, performed a post target turn to a withdrawal route which it held in common with the wave which had come in from the east. This resulted in each cell of the high wave being nearly superimposed over its counterpart cell during the withdrawal phase. As a result of the crossing track maneuver and the maneuver just described, the altitude restrictions imposed on the force were, as on the day before, critical.

The combination of the chaff seeded by the F-4s, mutual ECM support provided by nine B-52s in close proximity, a consolidated point attack from three widely separated axes of attack, and the varied post-target maneuvers performed by each wave added up to maximum ordnance on target in minimum exposure time, and confronted the defenses in one localized area with an impossible situation.

The frag order called for the same weight of effort as on the night before—a total of 60 B-52s. U-Tapao again provided 30, but the Andersen force was varied slightly to put 12 G models and 18 Ds in the North. Total bombing activity was rounded out by sending 30 G models on ARC LIGHT strikes in southern NVN and South Vietnam.[40]

The briefings for the last missions, while they lacked the high drama of the first briefings, still provided their own unique part of the story. General McCarthy recalls giving that last briefing:

As the crews filed into the briefing room, I could sense their rising level of confidence. We were closing in for the finale, and they knew it. The rumor had started floating around that this might be the last day of the big raids and they wanted to be a part of it. I had crews who had just landed hours earlier from the previous night's mission ask to be put in the lineup. Crews who had been designated as spares argued emphatically as to why they should be designated as primary crews, rather than spare.

One crew even went so far as to file an Inspector General complaint. Their argument was that they, being a less experienced crew, needed the mission for crew proficiency more than the older heads. With morale like that, I knew I had the best outfit in the United States Air Force. I knew how they felt. I had asked General Anderson to let me fly as the ABC, but he turned me down emphatically. I think the Flight Surgeon had said a few words to him in private. His compromise was to allow me the final briefing.

The launch went flawlessly. By now, the launch of a 30-plane mission had become a rather routine affair, but there was something about them that always drew a crowd. For this launch, there must have been at least 8,000 spectators along the flight line and gathered at vantage points on buildings. Sensing the end, offices all over the base closed down to let their people see it.[41]

LINEBACKER II | A VIEW FROM THE ROCK

29 DECEMBER 1972

B-52 CELLS/TARGET TIMES

'D' GUAM		'G' GUAM		'D' U-TAPAO	
AQUA	2320	PAINT	2320	GRAPE	2320
WALNUT	2323	BLACK	2322	MAPLE	2323
WINE	2326	LEMON	2324	CHESTNUT	2326
RED	2320	CHERRY	2338	OPAL	2336
RAINBOW	2323			BEIGE	2338
GREEN	2326			IVORY	2340
				TOPAZ	2342
				GRAY	2344
				CHROME	2334
				CINNAMON	2336

LEGEND

- - - - - - - CHINESE BUFFER ZONE
△ APPROXIMATE SAM COVERAGE
TARGETS
→ BOMBER ROUTE IN
→ BOMBER ROUTE OUT
COLOR   CALL SIGN OF CELL

TARGETS

1 PHUC YEN SAM SUPPORT   27
2 LANG DANG RAILROAD   18
3 TRAI CA SAM STORAGE   15
                          60

102 SUPPORT AIRCRAFT

EB-66 & EA-6B (NAVY) ECM
F-4 CHAFF
F-4 CHAFF ESCORT
F-4 (AF & NAVY) MIG CAP
F-4, B-52 ESCORT
F-105 & A-7 (NAVY) IRON HAND
F-4 HUNTER/KILLER

CHAPTER 6 | ACT THREE

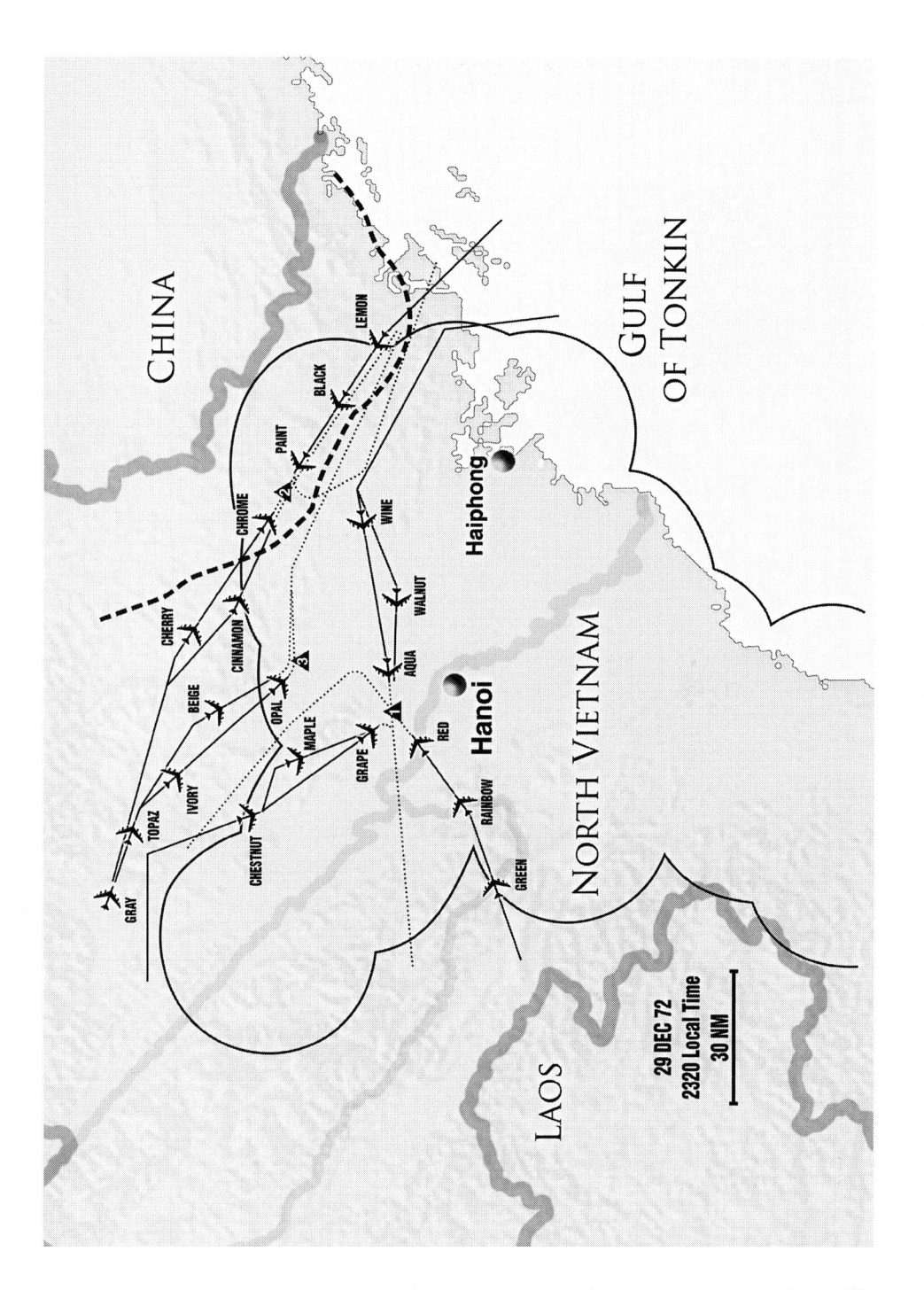

LINEBACKER II | A VIEW FROM THE ROCK

Mission execution on this last day went as directed, with one minor deviation which gave the opportunity to put to combat use one of the evolved tactics. Wine 3, the aircraft identified much earlier in this book as having a completely inoperative refueling system, had to return to the Rock. The other two Wine aircraft attached themselves to Walnut Cell just ahead, and struck Phuc Yen as a five-ship cell. That was an occasion for performing the SAM evasive maneuver with five aircraft in formation. No problems were reported with this tactic.[42]

No aircraft were lost or damaged for the second night in a row, and the few SAMs which were fired were even more erratic than on the previous night. One MIG pilot made a half-hearted attempt to engage Aqua Cell, but he was unsuccessful.

The bomb train from Gray 3 hit the Trai Ca SAM Storage Area, 40 miles north of Hanoi, at 17 minutes before midnight. The last Stratofortress bomb had fallen north of the 20th parallel. Minus Wine 3, the entire force went "over the fence" for the final time with 59.

Shortly after noon on December 30th, the last G model touched down on the Rock. LINEBACKER II was over.

Looking deceptively short through the telephoto lens, a B-52G deploys its drag chute during the landing roll.

CHAPTER 6 | ACT THREE

NOTES

1. *8AF History, V II*, p. 379.
2. *Chronology*, p. 122.
3. *USAF AIROPS*, p. IV-335.
4. Recollections of Author McCarthy, Blytheville AFB, AR., January 1978.
5. Lou Drendel, *B-52 Stratofortress in Action*, Warren, MI, Squadron/Signal Publications, Inc., 1975, pp. 39-40. Used with permission.
6. *43SW History*, pp. 99-100.
7. John W. Finney, "Attacks in North Redeem the B-52," the *Omaha World-Herald*, 24 December 1972, p. 1. See also John L. Frisbee, "The B-52: The Phoenix That Never Was," *Air Force Magazine*, February 1973, p. 4, and" As The Air War Hit A Peak," *U.S. News and World Report*, 8 January 1973, p. 4.
8. Recollection by Author Allison, Blytheville AFB, AR., autumn 1977.
9. January 1973 comment from a "Day 8" pilot, as reconstructed by author Allison, Blytheville AFB, AR., autumn 1977.
10. Recollection of Author McCarthy, Blytheville AFB, Ark., autumn 1977. For data on the Ebony 2 explosion, see: *Damage Analysis*, pp. A-27 to A-28. For data on the simultaneous damage to Cream 1 and 2 by a single SAM, see: *8AF History, V. II*, pp. 380-381. For data on the crash of Ash 1, see: *Damage Analysis*, pp. A-5 to A-6.
11. *SAC Participation*, pp. H-1 to H-3.
12. *USAF AIROPS*, pp. IV-197 to IV-200, and p. IV-277 contain details on Hunter/Killer concepts of operation. Note also p. IV-290 has weight of effort which varies with this present text.
13. *Tanker LINEBACKER II Chronology, 15-30 December 1972*, Report prepared by HQ SAC/DOTK, Offutt AFB, NE, 20 February 1973. SECRET. See also *SAC Participation*, pp. J-1 and J-2.
14. Colonel William W. Conlee, narrative written to authors, 12 May 1977.
15. *8AF History, V. II*, p. 380.
16. *SAC Participation*, p. K-1.
17. *8AF History, V. II*, p. 383.
18. *Ibid.*, p. 385.
19. *USAF in SEA*, p. 80.
20. *USAF AIROPS*, pp. IV-274 to IV-275.
21. *8AF History, V. II*, p. 523.
22. Lt Colonel Phillip R. Blaufuss, oral narrative to Lt Col Allison, 22 September 1977.
23. *43SW History*, p. 100.
24. *8AF History, V. II*, pp. 382 and 388.
25. *Damage Analysis*, p. A-11.
26. *Ibid.*, pp. A-21 to A-22.
27. Major Donald O. Aldridge, narrative written as Deputy Airborne Mission Commander, 27 December 1972, on file at A. F., Simpson Historical Research Center, Maxwell AFB, AL. SECRET
28. *Damage Analysis*, pp. A-97 and A-135.

29 *8AF History, V. II,* p. 390. 30. Ibid., p. 390.
30 *Ibid.,* p. 390.
31 *Hearings Before Sub-Committees of the Committee on Appropriations,* House of Representatives, 93d Congress (Tuesday, 18 January 1973), Washington, GPO, 1973, p. 4.
32 *USAF AIROPS,* p. IV-173.
33 *Chronology,* p. 277.
34 Notes by Author McCarthy, Blytheville AFB, AR., autumn 1977.
35 *8AF History, V. II,* p. 390.
36 *Chronology,* p. 322.
37 *USAF AIROPS,* p. IV-247.
38 *Ibid.,* pp. IV-315 to IV-316.
39 *8AF History, V. II,* p. 395.
40 *Ibid.,* p. 392.
41 Notes by Author McCarthy, Blytheville AFB, AR., autumn 1977.
42 *8AF History, V. II,* p. 393.

CHAPTER 7 | POSTLUDE
BUSINESS AS USUAL

By coincidence, Bob Hope and his troupe were at the base on December 30th on the homeward leg of his last overseas circuit, culminating a grand gesture of the human spirit.[1] His weary troupe was playing to an equally weary audience, which was reacting subconsciously to the experiences of an unprecedented two weeks. Four times Bob attempted to end his show, but each time the audience cheered the cast back for encores.[2] It is refreshing to speculate on what sort of gala reception his show would have received if the men and women of the Rock had known fully what that particular day meant to them and to the country at large. They would most assuredly have given him an unforgettable memory to be thankful for.

The campaign was over and national purpose had been served. Even so, as in so many instances of warfare, the smell of an intense struggle lingered over the scene. As the imaginary smoke of battle, which had been so symbolically portrayed by the black exhaust of hundreds of takeoffs, drifted out to sea, it was replaced by the exhaust of yet more launches. The war was still there.

No one seriously expected the war to disappear on the spot, much less the B-52 involvement in it. Still there was the vague uneasiness of "Where do we go from here?" To the majority at Andersen, the end of the up-country raids was not clear-cut. Wars aren't predictable, and parts of NVN were still being targeted.[3] The assurance that their performance in the northern heartland would not have to be repeated was not sealed until the signing of the Paris peace accords, establishing a Vietnam-wide cease-fire effective on 27 January 1973.[4]

That long-awaited news prompted a handwritten notice to be posted on the door of Charlie Tower: "This Property For Sale—On Or About 27 January."

Activities, however, did not slacken either as LINEBACKER II ended or the ceasefire was declared. The only people packing suitcases on the Rock and heading for the States were those involved in the normal TDY rotations.

Pressures for interdiction bombing in Laos and Cambodia escalated nearly in proportion to the reduction caused by the cease-fire, and Andersen remained a beehive of activity well into the new year. The continuation of bombing operations was scarcely distinguishable from

LINEBACKER II | A VIEW FROM THE ROCK

that of the pre-LINEBACKER II period.[5] There was some drawdown of the force, but the last B-52 mission in Southeast Asia was not flown until 15 August 1973. Many of the Eleven Day War flyers and their associates did not return home from their final ARC LIGHT/BULLET SHOT TDY until autumn.

Taking its cue from its KC-135 associate and its smaller fighter companions in the war, a 43d Strategic Wing BUFF smiles for the photographer following a combat mission.

The war was also not quite over for the POWs in Hanoi. Col Conlee concludes a story which began over the Bac Mai Storage Area on December 21st:

In late January 1973, everyone in our room was moved by enclosed jeep to a camp called the "Zoo." During this ride we saw some of the bombing effects, especially in the Hanoi Railyard, and were happily impressed. In the Zoo, most of the B-52 prisoners were put in four rooms in a rear compound called the "Pigsty" by the POWs. During this time our captors went to great lengths to prevent communication between camp areas. In these efforts, however, they were unsuccessful.

Shortly after our arrival at the Zoo the ceasefire occurred, and our conditions immediately improved. We were given two daily exercise breaks outside, more food to eat, and were not harassed as badly by the North Vietnamese. However, two incidents of ground-up light bulb glass in the soup occurred. We threatened a hunger strike, and the Vietnamese took precautions to stop this type of harassment. As our release date neared, the attempts to

CHAPTER 7 | POSTLUDE

force us into propaganda interviews and photos were stepped up, but we resisted. When a Soviet journalist was brought into the Pigsty area, we turned our backs on him, and he got disgusted and went elsewhere.

Release at Gia Lam Airport went smoothly. The Vietnamese were perturbed by our military bearing and the strict attention to detail which we all took to let them know they had not touched our spirit. The welcome at Clark Air Base was really overwhelming to all of us, especially since we had no news of how the POW returns had gone. The sight of real honest-to-goodness food was one of the most welcome sights at Clark, as were all the friendly faces. In my case, the warmth of the Texas welcome home at Sheppard AFB and again at Fort Worth and Dallas were memories of the best kind.[6]

SUMMARY

In the space of 11 days, B-52 Stratofortresses flew 729 sorties against 34 targets in North Vietnam above the 20th parallel. They expended over 15,000 tons of ordnance in the process. Bomb damage assessment revealed 1600 military structures damaged or destroyed, 500 rail interdictions, 372 pieces of rolling stock damaged or destroyed, three million gallons of petroleum products destroyed (estimated to be one-fourth of North Vietnam's reserves), ten interdictions of airfield runways and ramps, an estimated 80 percent of electrical power production capability destroyed, and numerous instances of specialized damage, such as to open storage stockpiles, missile launchers, and so forth.[7] No specific measurements are known of indirect losses, such as industrial inactivity, disruptions to almost all forms of surface travel, and communications outages.[8]

However, insight as to the indirect effects on all areas of the nation's productivity may be had by comparing one revealing statistic. Although the blockade of Haiphong harbor was in effect when LINEBACKER II started, logistic inputs to North Vietnam were assessed at 160,000 tons per month. In January 1973, imports dropped to 30,000 tons per month.[9]

In an attempt to defend against this destruction, the North Vietnamese used AAA, MIGs, and SAMs. No estimate of the amount of AAA expended is available, but its general persistence and intensity throughout the campaign suggests an enormous outlay of ammunition. The amount of cannon and rocket ordnance used by the MIGs against the force is likewise unknown, although it was definitely established that several active intercepts were attempted. The only stated figure indicative of the defensive reaction is a best estimate on total missiles fired at B-52s by the SAM sites—884 of them (one source suggests as many as 1,242 missiles; another estimates 914).[10] Of the conservative figure of 884, only 24 achieved hits, for a 2.7 percent success rate of launches to hits. Of the 24, only 15 resulted in a downed aircraft, one of which came within a breath of landing safely. Those equated to a 1.7 percent kill rate for the number of SAMs launched.

One aircraft was very slightly damaged by AAA. The NVN defenses hit 25 aircraft and downed 15. This represents 3.4 percent of the sorties hit, and 2.06 percent lost. Statistics relating to battle damage are subject to further interpretation, considering that one SAM detonation was credited with damaging two aircraft.

Of the total sortie count of 729 B-52s, 498 penetrated the especially high threat zones immediately surrounding Hanoi and Haiphong. These aircraft experienced a 4 percent loss rate.

There were 92 crewmembers aboard the 15 aircraft which went down. Of these, 61 were involved in the 10 B-52 losses over North Vietnam. At this writing, official Air Force reports show 14 known or declared KIA, while 14 remain MIA. Thirty-three who became POWs were subsequently repatriated.

Of the 31 crewmembers not downed over North Vietnam, 24 bailed out over Laos or Thailand and were rescued. The 25th man of this group was not known to bail out, but is presently still listed as MIA in Laos. The remaining six men were involved in a crash at U-Tapao, with four fatalities and two rescued.

ASSESSMENT

The fact that an aircraft which was designed in 1949 and first flown in 1952 could successfully penetrate a highly sophisticated air defense system, such as existed around Hanoi, is indicative of the quality of the airmen who maintained and flew it. This achievement further revealed a mixed force technological capability to stay abreast of the times, a capability whose limits were even then being sorely stretched by the aerodynamic and structural characteristics of the B-52. It is worth noting, more than five years later, that these same aircraft, in some cases older than the crewmembers who fly them, are still the only heavy bombers in our strategic arsenal.

When the 11 days of LINEBACKER II are viewed as a whole, one notable achievement of the campaign was the rapid change of complex tactics. Equally notable was the ability of the flight crews to be able to fly these complex tactics in combat, in mass formations, without benefit of practice. In the authors' opinion; this will eventually be recognized as one of the most outstanding feats of airmanship in strategic bombing operations in the history of aerial warfare.

The main complaint from these talented flyers and staff as LINEBACKER II ended was, "Why are we stopping now? With their defenses nearly in shambles, we could strike with impunity." Another week of these missions, they argued, and the North Vietnamese would have been suing for peace on our terms. There was some speculation later that, indeed, the North Vietnamese were prepared to sue for peace and cease all hostile activity in South

CHAPTER 7 | POSTLUDE

Vietnam, had LINEBACKER II continued. Some day perhaps this speculation may be verified. In the interim, one must ponder the analysis of a noted British expert on Southeast Asian wars, Sir Robert Thompson. In the book *The Lessons of Vietnam* he is quoted thus:

> In my view, on December 30, 1972, after eleven days of those B-52 attacks on the Hanoi area, you had won the war. It was over! They had fired 1,242 SAMs; they had none left, and what would come in overland from China would be a mere trickle. They and their whole rear base at that point were at your mercy. They would have taken any terms. And that is why, of course, you actually got a peace agreement in January, which you had not been able to get in October.[11]

The question often asked is "Was LINEBACKER II worth the losses in aircraft and crews?" Any loss, especially that of life, is grievous. However, it is the price of war. Rather than view the concept of loss in the negative, we point to the fact that B-52 losses were far lower than had been anticipated.[12] When compared to losses on prior mass bombardment missions where heavy enemy defenses were encountered, these losses were considerably less. All past indicators and the existing enemy threat suggested a loss rate of three to five percent. Many individual predictions doubled those percentages. The 15 B-52s lost during the 729 sorties flown resulted in a two percent loss rate, as has been noted. To put that figure in proper perspective, 22 aircraft would have to have been lost to increase the figure to three percent; 36 airplanes and crews would have increased it to five percent. Nevertheless, the latter two rates were viewed as "acceptable" when the battle was joined. Clearly, we would have lost more, had our crews not been the highly motivated, disciplined professionals they were, and had they not been well-supported in a fine combined effort.

The authors likewise argue that the losses were morally worth it. Rather than construct an elaborate defense of that position, we suggest careful consideration of only one comment. It is a quote from one of only a handful of people in the world whom we perceive to be authorities on the phenomenon known as peace, as it related to the Southeast Asian conflict. Most writings on the LINEBACKER II theme include the quote. Made by Dr. Henry Kissinger on January 24, 1973, it is here reiterated:

> . . . there was a deadlock . . . in the middle of December, and there was a rapid movement when negotiations resumed on January 8. These facts have to be analyzed by each person for himself . . .[13]

Three days after the diplomatic wording of that press conference, an agreed-upon cease-fire went into effect.

The authors have accepted Dr. Kissinger's challenge and made their analysis. We can't "prove" the validity of our decision. Someone, somewhere, in Vietnam may have the proof. We do know that, for 11 tense days and nights, our nation's tactical and strategic air forces delivered a blunt, emphatic message to Hanoi. The exclamation point at the end of that message was burned deeper and deeper in the minds of the North Vietnamese every time a B-52 radar navigator called "Bombs Away!"

The effect of the raids on the North Vietnamese prison guards at the Hanoi Hilton may indicate the psychological impact of the raids on those responsible for national policy. This impact was illustrated through information given the press and to the authors by former POWs.

Colonel Bill Conlee, recalling the ashen-faced Vietnamese who confronted him, stated:

There is no doubt in my mind but that LINEBACKER II was the primary reason for the negotiation decision by the North Vietnamese. They truly respect strength and, as seen up close, were absolutely terrified by the December 1972 B-52 bombing. I believe that this time period was certainly one of SAC's finest hours.[14]

Another POW related that anytime there was a raid on Hanoi, many of the prisoners would try to look out of their cell windows. Because most were high windows with bars, it was necessary for the prisoners to grasp the bars with their hands and lift themselves up to see out. When the guards saw the prisoners' hands on the bars, they would try to smash the prisoners' fingers with rifle butts. When the guards heard the long strings of B-52 bombs going off nearby, there was no such reaction. The POW reported he saw a guard, trembling like a leaf, drop his rifle and wet his pants.

Navy Captain Howard Rutledge is an advocate of the positive effects of the raids. A POW for over seven years, he told the story of a guard nicknamed "Parrot." Parrot was convinced by his own country's propaganda that all of Hanoi, including the Hanoi Hilton, was going to be wiped out. The POWs reassured him that, because of the skill of U.S. aviators, the Hilton was the truly safe place to be. Not yet convinced, he went outside to search for a haven—if one existed. Upon personally seeing the degree of destruction in specific target areas, Parrot returned to the Hilton and announced his intention not to leave again. The POWs comforted him thus: "Don't worry. Stay with us. We'll protect you."[15]

Colonel Robinson Risner, who spent seven and one-half years in prison, said he believed the release of American POWs came about largely because of President Nixon's decision to step up bombing and the introduction of B52 raids against the Hanoi-Haiphong area. Recalling the B-52 raids, Risner said:

On the 18th of December—I think that was the first night of the B-52 raids—there was never such joy seen in our camp before. There were people jumping up and down and putting their arms around each other and there were tears running down our faces.

We knew they were B52s and that President Nixon was keeping his word and that the Communists were getting the message.

CHAPTER 7 | POSTLUDE

We saw reaction in the Vietnamese that we had never seen under the attacks from fighters. They at last knew that we had some weapons they had not felt, and that President Nixon was willing to use those weapons in order to get us out of Vietnam.[16]

Finally, we cite our senior POW, Colonel (now Lieutenant General) John P. Flynn:

When I heard the B-52 bombs go off, I sent a message to our people. It said, "Pack your bags—I don't know when we're going home—but we're going home."[17]

Was it worth it? If LINEBACKER II was in any way responsible for validating national intent and for bringing our POWs back home, then we believe it was worth it.

NOTES

1. *8AF History, VII*, p. 586.
2. *303CAMW History*, unnumbered pages of photographs between pp. 8 and 9.
3. *USAF AIROPS*, p. IV-318.
4. *Ibid.*, pp. IV-326 to IV-327.
5. *8AF History, V I*, p. 7.
6. Colonel William W. Conlee, narrative written to authors, 12 May 1977.
7. Carl Berger *et al* ., *The United States Air Force in Southeast Asia 1961-1973*, Office of Air Force History, Washington, GPO, 1977, p. 166.
8. *USAF AIROPS*, pp. IV-313 to IV-314.
9. *Hearings Before Sub-Committees of the Committee on Appropriations*, House of Representatives, 93d Congress (Tuesday, 18 January 1973), Washington, GPO, 1973, p. 43.
10. Message: "Summary of SAM Firings," CINCPACAF/INK, 10/0130Z February 1973. SECRET
11. W. Scott Thompson and Donaldson D. Frizzell, ed., *The Lessons of Vietnam*, New York, NY, Crane, Russak & Co., 1977, p. 105. See also pp. 143, 168-170, and 177. Used with permission.
12. *8AF History, V. II*, pp. 344-345.
13. Henry A. Kissinger, News Conference on 24 January 1973. *The State Department Bulletin*, Volume LXVIII, Number 1755, 12 February 1973.
14. Colonel William W. Conlee, *op cit.*
15. Captain (USN) Howard E. Rutledge, Conversation with Lt Col Allison, 3 October 1977. See also Capt Rutledge's letter to the editor, "A POW View of LINEBACKER II," *Armed Forces Journal International*, September 1977, p. 20.
16. "Returned POW: 'Antiwar Crowd Kept Us in Prison,'" the *Omaha World-Herald*, 27 February 1973.
17. Lt General John P. Flynn, narrative written to authors, 3 August 1977.

APPENDIX

ORGANIZATIONS AND COMMANDERS, 1963-1974

8th Air Force
Moved, without personnel or equipment, from Westover AFB, MA, to Andersen AFB, Guam, on 1 April 1970. Replaced 3d Air Division. Moved, without personnel or equipment, to Barksdale AFB, LA, on 1 January 1975.

Lt Gen Alvan C. Gillem, II	1 Apr 1970	–	12 Jul 1970
Brig Gen Leo C. Lewis	13 Jul 1970	–	31 Jul 1970
Lt Gen Sam J. Byerley	1 Aug 1970	–	13 Sep 1971
Lt Gen Gerald W. Johnson	14 Sep 1971	–	30 Sep 1973
Lt Gen George H. McKee	1 Oct 1973	–	29 Aug 1974

3d Air Division
Reactivated at Andersen AFB, Guam, on 18 June 1954. Inactivated 31 March 1970. Replaced by 8 AF. Reactivated at Andersen AFB, Guam, on 1 January 1975.

Brig Gen Harold W. Ohlke	2 Jul 1963	–	16 Jul 1965
Maj Gen William J. Crumm*	16 Jul 1965	–	7 Jul 1967
Maj Gen Selmon W. Wells	8 Jul 1967	–	5 Jun 1968
Lt Gen Alvan C. Gillem, II	6 Jun 1968	–	31 Mar 1970

Air Division Provisional, 17
Activated 1 June 1972 at U-Tapao AB, Thailand, attached to 8 AF (SAC). Inactivated 1 January 1975.

Brig Gen Frank W. Elliott, Jr.	1 Jun 1972	–	5 Jun 1972
Brig Gen Glen R. Sullivan	6 Jun 1972	–	1 Feb 1973
Brig Gen Billy J. Ellis	2 Feb 1973	–	12 Oct 1973
Brig Gen James S. Murphy	13 Oct 1973	–	Aug 1974

General Crumm was killed in a B-52 mid-air collision on 7 Jul 1967.

Air Division Provisional, 57
Activated 1 June 1972 at Andersen AFB, Guam, attached to 8 AF (SAC). Inactivated 15 November 1973.

Brig Gen Andrew B. Anderson, Jr.	1 Jun 1972	– 14 Jan 1973
Brig Gen John W. Burkhart	15 Jan 1973	– Oct 1973
Brig Gen Edgar S. Harris, Jr.	Oct 1973	– 15 Nov 1973

43d Strategic Wing
Activated 1 April 1970 at Andersen AFB, Guam, assigned to 8 AF (SAC) and later attached to Air Division Provisional, 57 during existence of latter. Replaced 3960th Strategic Wing.

Col Lawrence E. Stephens	1 Apr 1970	– 30 Jun 1970
Col Glenn R. Dunlap	1 Jul 1970	– 28 Apr 1972
Col William P. Armstrong	28 Apr 1972	– 14 Jun 1972
Col James H. McGrath	16 Jun 1972	– 30 Nov 1972
Col James R. McCarthy	1 Dec 1972	– 1 Jun 1973
Col Morris E. Shiver	1 Jun 1973	– Aug 1973
Col Lawton W. Magee	Aug 1973	– 15 Nov 1973
Col James R. McCarthy	15 Nov 1973	– 15 Jun 1974

Strategic Wing Provisional, 72
Activated 1 June 1972 at Andersen AFB, Guam, attached to Air Division Provisional, 57. Inactivated 15 November 1973.

Col Kenneth M. Holloway	1 Jun 1972	– 8 Oct 1972
Col Thomas F. Rew	8 Oct 1972	– 16 Mar 1973
Col Thomas W. Sherman, Jr.	16 Mar 1973	– 16 Apr 1973
Col Nathaniel A. Gallagher	16 Apr 1973	– 15 Nov 1973

Consolidated Aircraft Maintenance Wing Provisional, 303
Activated 1 July 1972 at Andersen AFB, Guam, attached to Air Division Provisional, 57. Inactivated 15 November 1973.

Col Michael Perrone	15 Jun 1972	– 6 Aug 1972
Col James D. Naler	7 Aug 1972	– 31 Oct 1972
Col Thomas M. Ryan, Jr.	1 Nov 1972	– 1 Jun 1973
Col James R. McCarthy	1 Jun 1973	– 15 Nov 1973

APPENDIX

307th Strategic Wing
Activated 1 April 1970 at U-Tapao AB, Thailand, assigned to 8 AF (SAC) and attached to Air Division Provisional, 17 while the latter was in existence. Replaced 4258th Strategic Wing. Inactivated 30 September 1975.

Brig Gen Woodrow A. Abbott	1 Apr 1970	– 4 Jul 1970
Brig Gen John R. Hinton, Jr.	5 Jul 1970	– 4 Jul 1971
Brig Gen Frank W. Elliott, Jr.	5 Jul 1971	– 31 May 1972
Col Donald M. Davis	1 Jun 1972	– 10 Feb 1973
Col Bill V. Brown	11 Feb 1973	– 20 Aug 1973
Col Frank J. Apel, Jr.	21 Aug 1973	– 9 Apr 1974

Strategic Wing Provisional, 310
Activated 1 June 1972 at U-Tapao AB, Thailand, attached to Air Division Provisional, 17. Inactivated 1 July 1974.

Col James R. McCarthy	1 Jun 1972	– 13 Jun 1972
Col William L. Nicholson, III	14 Jun 1972	– 5 Dec 1972
Col Stanley C. Beck	6 Dec 1972	– 12 Jun 1973
Col Robert T. Herres	13 Jun 1973	– 7 Sep 1973
Col Vernon R. Huber	15 Sep 1973	– 20 Nov 1973
Col Richard J. Smith	21 Nov 1973	– 1 Jan 1974

Consolidated Aircraft Maintenance Wing Provisional, 340
Activated 1 July 1972 at U-Tapao AB, Thailand, attached to Air Division Provisional, 17. Inactivated 1 July 1974.

Col Marvin L. Adams	1 Jul 1972	– 7 Aug 1972
Col Michael Perrone	7 Aug 1972	– 16 Dec 1972
Col William B. Maxson	16 Dec 1972	– 7 Jun 1973
Col Melbourne Kimsey	8 Jun 1973	– 28 Oct 1973
Col Joseph P. Cerny	29 Oct 1973	– 14 Nov 1973
Col Jerome F. O'Malley	15 Nov 1973	– Jan 1974

376th Strategic Wing
Activated 1 April 1970 at Kadena AB, Okinawa. Replaced 4252d Strategic Wing.

Brig Gen Alan C. Edmunds	1 Apr 1970	– 1 Sep 1970
Col Jack A. Weyant	2 Sep 1970	– 30 Aug 1972
Col Dudley G. Kavanaugh	30 Aug 1972	– Sep 1974

3960th Strategic Wing

Activated at Andersen AFB, Guam, 1 April 1955, assigned to 3d Air Division. Underwent short-lived changes of designation to Air Base Wing and Combat Support Group. Inactivated 31 March 1970, replaced by 43d Strategic Wing.

Col Edward C. Unger	Apr 1964	–	21 Jul 1964
Col Edward D. Gaitley, Jr.	22 Jul 1964	–	9 Jul 1965
Col Joseph J. Semanek	10 Jul 1965	–	11 Jul 1967
Col James M. Smith	12 Jul 1967	–	6 Jul 1969
Col Lawrence E. Stephens	7 Jul 1969	–	31 Mar 1970

Bombardment Wing Provisional, 4133

Activated at Andersen AFB, Guam, 1 February 1966. Turned over its combat mission to 43d Strategic Wing and inactivated on 1 July 1970.

Col William T. Cumiskey	1 Feb 1966	–	31 Mar 1966
Col Harold J. Whiteman	1 Apr 1966	–	12 Jun 1966
Col Albert H. Schneider	13 Jun 1966	–	20 Sep 1966
Col Willard A. Beauchamp	21 Sep 1966	–	27 Sep 1966
Col Earl L. Johnson	28 Sep 1966	–	28 Feb 1967
Col Mitchell A. Cobeaga	1 Mar 1967	–	30 Oct 1967
Col Robert E. Brofft	31 Oct 1967	–	28 Mar 1968
Col Madison M. McBrayer	29 Mar 1968	–	31 Aug 1968
Col Robert E. Blauw	1 Sep 1968	–	25 Sep 1968
Col Robert E. Brofft	26 Sep 1968	–	20 Mar 1969
Col Robert E. Blauw	21 Mar 1969	–	19 Sep 1969
Col Raymond P. Lowman	20 Sep 1969	–	29 Sep 1969
Col Howard P. McClain	30 Sep 1969	–	24 Mar 1970
Col Harold E. Ottoway	25 Mar 1970	–	Jun 1970
Col William P. Armstrong	Jun 1970	–	1 Jul 1970

4252d Strategic Wing

Organized and activated 12 January 1965 at Kadena AB, Okinawa. Inactivated 1 April 1970. Replaced by 376th Strategic Wing.

Col Holley W. Anderson (Acting)	12 Jan 1965	–	17 Feb 1965
Col Morgan S. Tyler, Jr. (later BG)	18 Feb 1965	–	18 Jul 1967
Col Eugene A. Stalzer (later BG)	19 Jul 1967	–	3 Aug 1969
Brig Gen Alan C. Edmunds	4 Aug 1969	–	31 Mar 1970

APPENDIX

4258th Strategic Wing
Activated at U-Tapao AB, Thailand 2 June 1966 to 1 April 1970. Redesignated 307th Strategic Wing.

Capt Ralph W. Ingram	2 Jun 1966	– 20 Jul 1966
Col John W. Farrar	21 Jul 1966	– 30 Jun 1967
Col Alex W. Talmant	1 Jul 1967	– 4 Aug 1968
Brig Gen Richard M. Hoban	5 Aug 1968	– 10 Jul 1969
Brig Gen Woodrow A. Abbott	11 Jul 1969	– 31 Mar 1970

43d Combat Support Group
Activated 1 July 1970 at Andersen AFB, Guam, assigned to 43d Strategic Wing.

Col Arthur G. Ray	1 Jul 1970	– 11 Nov 1970
Col Edward M. McDonald	12 Nov 1970	– 15 Nov 1970
Col Dwayne E. Kelly	16 Nov 1970	– 1 Oct 1972
Col John H. Vincent	2 Oct 1972	– Jan 1974

Air Refueling Squadron Provisional, 4101
Activated at Takhli AB, Thailand, 6 June 1972, attached to Air Division Provisional, 17. Inactivated 15 February 1973.

Col John H. Moore	6 Jun 1972	– 1 Nov 1972
Col Rodger L. Brooks	2 Nov 1972	– 12 Feb 1973

Air Refueling Squadron Provisional, 4102
Activated at Clark AB, Republic of Philippines, 6 June 1972, attached to 376th Strategic Wing, relocated to Ching Chuan Kang AB, Taiwan, inactivated 8 November 1972. Reactivated at Clark AB, 18 December 1972, inactivated 22 January 1973,

Col Robert L. Nicholl	6 Jun 1972	– 25 Oct 1972
Col Jacques K. Tetrick	26 Oct 1972	– 8 Nov 1972
Col William L. Nicholson, III	18 Dec 1972	– 22 Jan 1973

Air Refueling Squadron Provisional, 4103
Activated at Don Muang Airport, Thailand, 1 July 1972, attached to Air Division Provisional, 17. Inactivated 10 October 1972.

Col William E. Long	1 Jul 1972	– 10 Oct 1972

Air Refueling Squadron Provisional, 4104
Activated 9 June 1972 at Korat AB, Thailand, attached to Air Division Provisional, 17. Inactivated 8 November 1972.

Col James R. McCarthy	9 Jun 1972	– 3 Sep 1972
Col Rodger L. Brooks	4 Sep 1972	– 8 Nov 1972

4220th Air Refueling Squadron
Activated at Ching Chuan Kang AB, Taiwan, 2 February 1968, assigned to 4252d Strategic Wing. Inactivated 31 January 1971.

Col Glen L. Pugmire	2 Feb 1968	– 19 Feb 1969
Col Robert L. Holladay	20 Feb 1969	– 17 May 1970
Col George O. Bolen	18 May 1970	– 7 Jul 1970
Col Lewis H. Richardson	7 Jul 1970	– 31 Jan 1971

GLOSSARY

ABBREVIATION	DESCRIPTION
A-7	"Corsair II," U.S. Navy fighter aircraft
AAA	Antiaircraft artillery
ABC	Airborne Mission Commander (also Airborne Commander)
AC	Aircraft Commander, the pilot responsible for the performance and operation of the aircrew and aircraft
AC	Alternating current
AD	Air division
AF	Air Force, in this case a Numbered Air Force (8 AF)
AFB	Air Force Base
Air abort	Cancellation of an aircraft mission for any reason other than enemy action, at any time, from takeoff to mission completion
Aircrew	The full complement of air officers and airmen who man, or are designated to man, an aircraft in the air; often shortened to "crew"
Andersen	U.S. Air Force Base at Guam, Mariana Islands
AQM-34L	"Firebee," low altitude reconnaissance drone
ARC LIGHT	Overall term for B-52 operations in Southeast Asia
B-52	"Stratofortress"; heavy jet bomber. Two model series, the "D" and "G," were used in LINEBACKER II operations
Ballistic	Unguided, i.e., follows a ballistic trajectory when thrust is terminated
BDA	Bomb damage assessment
Beeper	Emergency radio in crewmember's parachute pack, automatically activated upon ejection or bailout

ABBREVIATION	DESCRIPTION
Bicycle Works	Informal nickname given to the consolidated maintenance complex at Andersen AFB
BRL	Bomb release line, imaginary or theoretical line around a target area at which a bomber releases its first bomb. For B-52 ARC LIGHT operations, this line was reached approximately six miles prior to the target
BUFF	Informal nickname and acronym for the B-52, derived from "Big Ugly Fat Fella"
BUFFALO HUNTER	Nickname for DC-130/ AQM-34L launch reconnaissance and recovery operations involving drone photography and intelligence
BULLET SHOT	Nickname for the buildup of B-52 forces in Thailand and Guam to support Southeast Asian military operations in 1972 and 1973
Burn-through	The zone around a ground-based radar system where the power output exceeds the power of airborne jamming signals
BW	Bombardment Wing
CAMW	Consolidated Aircraft Maintenance Wing
CAP	Combat air patrol, an aircraft patrol provided to protect the force by intercepting and destroying hostile aircraft before they reach their intended target
Cell	For B-52 operations, a planned formation of three aircraft
Chaff	Lightweight strips of metal foil or fibreglass, cut to various lengths, packed in small bundles and dispensed in flight to break open and form "clouds" of reflective materials, thus interfering with radars operating at the same frequency length
Charlie Tower	Control tower for ground movement, launch, and recovery of aircraft at the two bases of Andersen and U-Tapao
CINCSAC	Commander-in-Chief, Strategic Air Command
Clock Position	Horizontal position of any object relative to the aircraft, with the aircraft nose representing 12 o'clock. A fighter at 6 o'clock would be directly behind
COA	Confirmed operational area for surface-to-air missiles
COL	Confirmed operational location, more specific than a COA
COMBAT APPLE	Nickname for reconnaissance operations performed by RC-135s

GLOSSARY

ABBREVIATION	DESCRIPTION
Compression	Three or more B-52 cells striking the same target area with up to ten minutes' separation between cells
Contrail	Condensation trail, a visible trail of water droplets or ice crystals sometimes forming in the wake of an aircraft
CONUS	Continental United States, the 48 contiguous states excluding Alaska and Hawaii
Cousin Fred	Supervisor of taxiing and towing of aircraft at Andersen AFB
CP	Copilot, coordinates flight control of aircraft with pilot, second in command
CTF-77	Commander, Task Force 77, responsible for U.S. Navy operations in the Gulf of Tonkin, Vietnam
DABC	Deputy Airborne Mission Commander
DC-130A	"Hercules," carrier aircraft for the AQM-34L low altitude reconnaissance drone
Decision Speed	The speed during takeoff roll at which a pilot must decide to continue with the takeoff or to abort
DMZ	Demilitarized zone, buffer zone between North and South Vietnam
DNIF	Duty not involving flying
DOWNLINK	Beacon signal transmitted from a surface-to-air missile to the command guidance radar site
DOX	Directorate of Operations Plans
Drag Chute	Deceleration parachute for aircraft
EA-3	"Skywarrior," U.S. Navy attack bomber modified for ECM operations
EA-6	"Intruder," U.S. Navy and Marine attack bomber modified for ECM operations
EB-57	"Canberra," U.S. Air Force version of British bomber, modified for electronic jamming and penetrating of air defenses
EB-66	"Destroyer," light jet bomber modified for ECM operations
EC-121	Early-warning, fighter-control, and reconnaissance aircraft derived from the Super Constellation transport

ABBREVIATION	DESCRIPTION
EC-135	Modified version of KC-135 Stratotanker, used in communications/command and control
ECM	Electronic countermeasures, any of various measures using electronic and reflecting devices to reduce the military effectiveness of enemy equipment or tactics employing or affected by electro-magnetic radiations
Eleven Day War	Frequently used nickname for LINEBACKER II
ELINT	Electronic intelligence
EW	Electronic Warfare Officer; responsible for identification, assessment, and defense against the whole spectrum of threats to a penetrating bomber
F-4	"Phantom II," U.S. Air Force and Navy tactical all-purpose fighter
F-105	"Thunderchief," fighter-bomber
F-111	Variable geometry (wing sweep) tactical fighter
Fansong	North Atlantic Treaty Organization (NATO) designator for ground-based guidance radar used with the SA-2 missile
FCS	Fire control system; automatic radar tracking or manually controlled, electrically/hydraulically powered turret containing four M-3 .50 caliber guns. Defends the B-52D and G in the aft quadrant
Feet Wet	Crossing hostile coastline outbound
Flak	Acronym for all forms of fragments of exploding munitions, especially that expended against aircraft
Flame(d) Out	The extinguishment of the flame in a reaction engine, especially a jet engine
Frag	Fragmentary order, a specific portion of the overall contingency operation, providing detailed guidance to a small segment of the force (in this case, for a cell)
"F" Troop	Originally a television series depicting the inept life and actions of a bumbling group of post-Civil War cavalrymen
G (AG)	Gunner, responsible for defense of B-52D and G from rearward aerial assault, using .50-caliber FCS
GIANT SCALE	Nickname for SR-71 high altitude reconnaissance operations

GLOSSARY

ABBREVIATION	DESCRIPTION
GPI Point	Ground position indicator, readily identified radar return of known geographic coordinates, used for navigation
Haiphong	Major coastal port city of North Vietnam
Hanger	A bomb which fails to release
HF	High frequency radio, long range
Hanoi	Capitol city of North Vietnam
HH-3E	"Jolly Green Giant," rescue helicopter developed originally to facilitate penetration deep into North Vietnam on rescue missions
HH-53B	"Super Jolly," helicopter designed to supplement the HH-3E mission (faster and larger)
Hunter/Killer	Term used to identify tactical operations against hostile defensive systems, primarily those employing radar guided antiaircraft ordnance. Conducted with various combinations of aircraft-one type identifying and attacking and others keying on the initial strike
IFR	Instrument flight rules; used also to describe weather conditions in which flight is conducted using only instruments as references
Iron Hand	Nickname for tactical operations to identify, locate, and suppress SAM defenses and radar-controlled AAA sites
IP	Initial point, a point on the ground (identified visually, by electronic means, or by navigation) over which an aircraft begins a bomb run
JCS	Joint Chiefs of Staff
JP-4	Designator for one type of jet propulsion fuel
KC-135	"Stratotanker," provides aerial refueling for all types of tactical and cargo aircraft. Also used as long-range passenger or cargo aircraft, or as both
Kadena	Air Base at Okinawa, Ryukyu Islands, Japan
KIA	Killed in action
LINEBACKER I	Nickname for overall interdiction bombing effort in North Vietnam during the spring, summer, and autumn of 1972. JCS-directed USAF strikes from 9 May 1972 to 22 October 1972
LINEBACKER II	Nickname for overall bombing effort in North Vietnam in December 1972. JCS-directed U.S. strikes from 18 to 29 December 1972

ABBREVIATION	DESCRIPTION
LSL	Lethal SAM line, identifies any point from a missile launch site where a combination of missile performance and command/guidance capability will result in a probable "hit"
M-117	750-pound class, general purpose, conventional bomb
MK-82	500-pound class, general purpose, conventional bomb
MIA	Missing in action
MIG	Name for the Mikoyan/Gurevich series of Soviet jet fighter aircraft
MIG-21	NATO designator "Fishbed," delta-wing fighter/interceptor
MIG CAP	Combat air patrol, fighters deployed between MIG threat and force to be protected. Commonly used to identify general fighter protection operations. Other terms are Barrier CAP (BARCAP) and MIGSCREEN
mm	Millimeter, as in 100-mm AAA
N (NAV)	Navigator, responsible for all general navigation requirements. Also responsible along with radar navigator for assuring accurate bomb delivery
NCO	Noncommissioned officer
NIKE	A U.S. Army, rocket-propelled, surface-to-air guided missile, without wings, using booster rockets
NKP	Nakhon Phanom Air Base, Thailand
Northwest Field	Inactive airfield at north end of Guam
NVN	Used interchangeably to identify North Vietnam and North Vietnamese
OAP	Offset aiming point, a precise radar return located a computed distance from a target, selected to assure target area identification and bombing accuracy
OLYMPIC TORCH	Nickname for U-2R high altitude reconnaissance operations
P	Pilot, used interchangeably with aircraft commander
P (Prov)	Provisional, a unit organized for a limited period of time
PCS	Permanent change of station, being "permanently" assigned to the base or installation

GLOSSARY

ABBREVIATION	DESCRIPTION
Ploesti	City and oil refinery in Rumania, attacked at low level by B-24s from Libyan bases in World War II on August 1, 1943. Bombed several times in that conflict
POL	Petroleum, oil, and lubricants
POW	Prisoner of war
PPS	Petroleum products storage
Press-on	A mission where aircraft continue their attack regardless of the nature or intensity of defensive reaction
PTT	Post-target turn
Pylon	A projection under an aircraft wing, originally designed on the B-52 to carry air-to-ground missiles but modified on the B-52D to suspend conventional ordnance-12 bombs under each wing
RC-135	Modified version of KC-135 Stratotanker, used for reconnaissance operations
Red Ball	Highest priority maintenance or delivery of aircraft parts, usually associated with timing limitations
Red Crown	Call sign of TF77 Navy vessel (specifically the USS Long Beach) in Gulf of Tonkin, provided coordination and control to friendly forces and monitored enemy air order of battle
Regensburg	Site of German Messerschmitt aircraft factories; successfully attacked in one of the first mass assault, long-range bombardment missions of World War II
RN	Radar Navigator, primarily responsible for bomb release, assists in navigation to and from the target
Rock, The	Nickname applied to the island of Guam, specifically Andersen AFB
ROLLING THUNDER	Nickname for protracted interdiction air campaign against North Vietnam, 1965-1968
SA-2	NATO designator "Guideline," Soviet-built surface-to-air missile, designed to counter the B-52 at high altitude, but used extensively against all U.S. aircraft in Southeast Asia
SAC	Strategic Air Command
SAM	Surface-to-air missile

ABBREVIATION	DESCRIPTION
SAR	Search and rescue
SCAT	SAC contingency aircrew training
Schweinfurt	Site of German ball-bearing factories crippled in two long-range bombardment missions which were instrumental in establishing daylight bombardment as an effective strategic weapon in World War II
SEA	Southeast Asia. Also sometimes referred to as SEAsia
17th Parallel	Line of North latitude used to establish a recall point for B-52 strike missions in LINEBACKER II
SHRIKE	AGM-45A supersonic air-to-ground missile designed to home automatically on enemy radar installations
SIGINT	Signal intelligence
Sortie	Flight or mission by an individual aircraft
SR-71	"Blackbird," high speed, high altitude reconnaissance and intelligence aircraft
Stream	A flow of aircraft following approximately the same route
Standdown	Term meaning that no aircraft fly, or that aircraft do not participate in specific air operations
SW	Strategic Wing
TAC	Tactical/Tactical Air Command
TACAIR	Tactical air
TDY	Temporary duty
Thud Ridge	Nickname for a mountain range beginning about 20 miles north-northwest of Hanoi and extending about 25 miles to the northwest; prominent feature on radar, used for navigation and target orientation
TOT	Time on target; also time over target. Time of impact of bomb(s) on the target site
Trail Formation	Aircraft directly behind one another
TTR	Target tracking radar; used to identify the maneuver flown to counter the threat and deny tracking information to the hostile radar
U-2R	High altitude reconnaissance and intelligence gathering aircraft

GLOSSARY

ABBREVIATION	DESCRIPTION
Uncle Ned	Maintenance coordinator in Charlie Tower
Uncle Tom	Coordinator in Charlie Tower for ground movement and location of aircraft
UPLINK	Command guidance signal transmitted from a ground radar site to a launched surface-to-air missile
USAF	United States Air Force
USN	United States Navy
USMC	United States Marine Corps
U-Tapao	Royal Thai Navy Airfield, south-southeast of Bangkok, Thailand, also called "U-T"
VN-xxx	Designation of a North Vietnamese missile launch site
Wave	A succession of aircraft formations which move across or against a target or other point
Wild Weasel	F-105F and F-4 aircraft with radar homing and warning equipment and antiradiation missiles, enabling them to home on SA-2 radar guidance signals and to mark the location of missile sites
Z (Zulu)	Greenwich Mean Time, the international time zone standard from which all local times may be computed. Used to avoid confusion in timing references

BIBLIOGRAPHY

Photographic documentation was obtained as indicated under "Acknowledgements." Most unit histories cited in the bibliography contain useful photographs. One history of particular value, not referred to under "Notes," contains both extensive bomb damage assessment photography and supporting summary charts and maps. See the bibliography entry, *History of The Air Force Intelligence Service*.

The diagram of the B-52G crew compartment which appears in Chapter II is an adaptation of an official USAF photo appearing in the book: Holder, William G., *Boeing B-52 "Stratofortress,"* Aero Series, Volume 24, Fallbrook, California, Aero Publishers, Inc., p. 91. It is used with permission. [The illustration in the 2018 Edition is a new version, by Zaur Eylanbekov.]

[Note to the 2018 edition: Classifications are listed in the notes and bibliography as received from the original edition. No attempt was made to update classification markings.]

B-52 Combat Damage Analysis, Prepared by the Caywood-Schiller Division of A.T. Kearney, Inc. for the Joint Technical Coordinating Group for Munitions Effectiveness, Published as 61 JTCG/ME-75-l, October 1974. SECRET

Berger, Carl *et al.*, *The United States Air Force in Southeast Asia 1961-1973*, Office of Air Force History, Washington, U.S. Government Printing Office, 1977.

Bomber ARC LIGHT Crew Manual, 8 AFM 55-2, HQ 8AF, Andersen AFB, Guam, Mariana Islands, 1 November 1972. SECRET

Chronology of SAC Participation in LINEBACKER II, HQ SAC/[HO], Offutt AFB, NE, 12 August 1973 (Also found as *USAF Air Operations in Southeast Asia, 1 July 1972 15 August 1973*, Volume V, CORONA HARVEST, Prepared by HQ PACAF with support of SAC). TOP SECRET

Coffey, Thomas M., *Decision Over Schweinfurt, The U.S. 8th Air Force Battle for Daylight Bombing*, New York, NY, David McKay Company, Inc., 1977.

COMMANDO HUNT V, Report prepared by HQ 7AF, Tan Son Nhut AB, South Vietnam, May 1971. SECRET

COMMANDO HUNT VII, Report prepared by HQ 7AF, Tan Son Nhut AB, South Vietnam, June 1972. SECRET

Crew Narrative of 23 November 1972 Combat Mission, Cassette tape on file at Alfred F. Simpson Historical Research Center, Maxwell AFB, AL. SECRET

Combat Mission Tape, Westover Crew E-12, 19 December 1972, Cassette tape on file at Alfred F. Simpson Historical Research Center, Maxwell AFB, AL.

Drendel, Lou, *B-52 Stratofortress in Action*, Warren, MI, Squadron/Signal Publications, Inc., 1975.

Dugan, James and Carroll Stewart, *Ploesti, The Great Ground-Air Battle of 1 August 1943*, New York, NY, Random House, 1962.

Flight Manual, T.O. 1B-52D-l, Air Logistics Command, Tinker AFB, OK, 1 August 1974.

Flight Manual, T.O. 1B-52G-l, Air Logistics Command, Tinker AFB, OK, 1 January 1975.

Futrell, R. Frank et al., *Aces and Aerial Victories: The United States Air Force in Southeast Asia 1965 - 1973*, The A.F. Simpson Historical Research Center, Maxwell AFB, AL, and Office of Air Force History, HQ USAF, 1976, Washington, GPO, 1977.

GIANT STRIDE VII, SAC OT&E Final Report, HQ SAC/[DOXT], Offutt AFB, NE, 31 August 1971. SECRET

Goldwater, Senator Barry M., "Airpower in Southeast Asia," *Congressional Record*, Volume 119, Part 5, 93d Congress, 1st Session, 26 February 1973.

Hearings Before Sub-Committees of the Committee on Appropriations, House of Representatives, 93d Congress (Tuesday, 18 January 1973), Washington, GPO, 1973.

History of Eighth Air Force, 1 July 1972 - 30 June 1973, Volume I of VIII Volumes, Narrative-Part I, Andersen AFB, Guam, M.I., 23 August 1974. SECRET

History of Eighth Air Force, 1 July 1972-30 June 1973, Volume II of VIII Volumes, Narrative-Part II, Andersen AFB, Guam, M.I., 23 August 1974. SECRET

History of 43rd Strategic Wing, 1 July 1972-31 December 1972, BULLET SHOT Part II, With Emphasis on LINEBACKER II, Volume I, Andersen AFB, Guam, M.I., 24 May 1973. SECRET

History of 72nd Strategic Wing (Provisional), 1 November 1972-31 January 1973, Volume I, Andersen AFB, Guam, M.I., 28 April 1973. SECRET

History of Strategic Air Command—FY 1973, Volume II, Narrative, Prepared by HQ SAC/HO, Offutt AFB, NE, 2 May 1974. SECRET

History of The Air Force Intelligence Service, 1 July 1972-30 June 1973, Volume III, Supporting Document K-2, *LINEBACKER II, Summary*. SECRET

BIBLIOGRAPHY

History of 303rd Consolidated Aircraft Maintenance Wing (Provisional), 1-31 December 1972, Andersen AFB, Guam, M.I., 20 January 1973. SECRET

History of 307th Strategic Wing, April-June 1972, Volume I, U-Tapao Royal Thai Navy Airfield, Thailand, 25 September 1972. SECRET

History of 307th Strategic Wing, October-December 1972, Volume I, U-Tapao Royal Thai Navy Airfield, Thailand, 12 July 1973. SECRET

History of 307th Strategic Wing, October-December 1972, Volume IV, Appendix U, Mission Charts, U-Tapao Royal Thai Navy Airfield, Thailand, 12 July 1973. SECRET

Kissinger, Henry A., News Conference on 24 January 1973, *The State Department Bulletin,* Volume LXVIII, Number 1755, 12 February 1973.

Kissinger, Henry A., "Vietnam Peace Negotiations," News Conference on 26 October 1972, *Weekly Compilation of Presidential Documents,* Volume 8, Number 44, 30 October 1972.

Message (TS), JCS to CINCPAC and CINCSAC, JCS 3348, for Gayler and Meyer from Moorer, 15/0147Z December 1972 (72-B-7576)

Message (TS), JCS to CINCPAC, JCS 5829, for Gayler from Moorer, 18/0015Z December 1972 (72-B-7620).

Message (TS), JCS to CINCPAC, CINCSAC, and COMUSMACV, JCS 7807, for Gayler, Meyer, and Weyand from Moorer, 19/2322Z December 1972 (72-B-7673).

Message (S-NOFORN-GDS-80), 7AF/SAC ADVON to 8AF *et al.,* "Current B-52 Tactics", 21/0830Z December 1972.

Message (S-GDS-80), 307SW/17AD/CC to 8AF/CC, "ARC LIGHT Compression Tactics," 22/0806Z December 1972.

Message, "Summary of SAM Firings," CINCPACAF/INK, 10/0130Z February 1973.

Omaha World-Herald, Published daily, Omaha, NE.

Pacific Daily News, Published daily, Agana, Guam, Mariana Islands.

SAC Operations in LINEBACKER II, Tactics and Analysis, Briefing prepared by HQ SAC/XOO, Offutt AFB, NE, 3 August 1976. SECRET

SAC Participation in LINEBACKER II, Volume I, Basic Report, HQ SAC/[XOO], Offutt AFB, NE, 5 January 1973. TOP SECRET

SAC Southeast Asia Progress Report #84, HQ SAC/[ACM], Offutt AFB, NE, November 1972. SECRET

Summary of Tactics, Report prepared by 8AF/IN, Andersen AFB, Guam, M.I., 1 February 1973. SECRET

Supplemental History on LINEBACKER II (18-29 December), 43rd Strategic Wing and Strategic Wing Provisional, 72nd (Volume I), Air Division Provisional, 57th, Eighth Air Force, Andersen AFB, Guam, M.I., 30 July 1973. TOP SECRET

Tanker LINEBACKER II Chronology, 15-30 December 1972, Report prepared by HQ SAC/DOTK, Offutt AFB, NE, 20 February 1973. SECRET

The New York Times, Published daily, New York, NY.

The USAF in Southeast Asia 1970-1973, Lessons Learned and Recommendations: A Compendium, CORONA HARVEST, Prepared by HQ PACAF, Hickam AFB, HI, 16 June 1975. SECRET

Thompson, W. Scott and Donaldson D. Frizzell, ed., *The Lessons of Vietnam,* New York, NY, Crane, Russak & Co., 1977.

USAF Air Operations in Southeast Asia, 1 July 1972-15 August 1973, Volume II, CORONA HARVEST, Prepared by HQ PACAF with support of SAC, Hickam AFB, HI, 7 May 1975. TOP SECRET

USAF Oral History Interview Program, with Lt General Gerald W. Johnson, By Charles K. Hopkins, 8AF Historian, Andersen AFB, Guam, M.I., 3 April 1973. SECRET

Whalen, Norman M., "Ploesti: Group Navigator's Eye View", *Aerospace Historian,* Volume 23, Number 1, Spring/March 1976, Manhattan, KS, Department of History, Kansas State University, 1976.

EXPANDED BIBLIOGRAPHY FOR THE 2018 EDITION

BOOKS

Michel, Marshall L. III. *America's Last Vietnam Battle: The 11 Days of Christmas.* San Francisco: Encounter Books, 2002.

—. *Clashes: Air Combat over North Vietnam, 1965-1972.* Annapolis: Naval Institute Press, 1997.

Head, William P. *War from above the Clouds: B-52 Operations during the Second Indochina War and the Effects of the Air War on Theory and Doctrine.* Maxwell AFB: Air University Press, 2002.

Clodfelter, Mark. *The Limits of Air Power: The American Bombing of North Vietnam.* New York: Free Press, 1989.

Eschmann, Karl J. *Linebacker: The Untold Story of the Air Raids over North Vietnam.* NewYork: Ivy Books, 1989.

Gilster, Herman L. *The Air War in Southeast Asia: Case Studies of Selected Campaigns.* Maxwell AFB: Air University Press, 1993.

McCarthy, James R. and Robert E. Rayfield. *B-52s over Hanoi: A Linebacker II Story.* Fullerton: California State Fullerton Press, 1996.

Sharp, Ulysses S. Grant. *Strategy for Defeat: Vietnam in Retrospect.* San Rafael, California: Presidio Press, 1978.

Summers, Harry G. *On Strategy: A Critical Analysis of the Vietnam War.* Novato, California: Presidio Press, 1995.

Thompson, Wayne. *To Hanoi and Back: The U.S. Air Force and North Vietnam, 1966-1973.* Washington: Smithsonian Institution Press, 2000.

PERIODICALS

Allison, George B. "The Bombers Go to Bullseye." *Aerospace Historian,* December 1982.

Ball, George W. "Top Secret: The Prophecy the President Rejected: How Valid Are the Assumptions Underlying Our Viet-Nam Policies?" *The Atlantic,* July 1972.

Boyne, Walter J. "Linebacker II." *Air Force,* November 1997.

Brownlow, Cecil. "North Viet Bombing Held Critical." *Aviation Week & Space Technology,* March 5, 1973.

Testimony of Admiral Moorer before the House Appropriations Subcommittee on Defense.

Burdick, Frank A. "The Christmas Bombing of North Vietnam: The Limitations of Force."

Journal of Political & Military Sociology 12, Fall 1984.

Drenkowski, Dana. "Operation Linebacker II, Part 1." *Soldier of Fortune,* September 1977.

—. "Operation Linebacker II, Part 2." *Soldier of Fortune,* November 1977.

—. "The Tragedy of Linebacker II." *Armed Forces Journal International* 114, July 1977.

See also: "Tragedy of Linebacker II: The USAF Response." *Armed Forces Journal International* 114, August 1977.

Drenkowski, Dana and Lester W. Grau. "Patterns and Predictability: The Soviet Evaluation of Operation Linebacker II." *Journal of Slavic Military Studies* 20, December 2007.

"Effects of the Bombing." *Time,* June 26, 1972.

Ellis, Richard H. and Frank B. Horton, III. "Flexibility—A State of Mind." *Strategic Review* 4, Winter 1976.

Ginsburgh, Robert N. "North Vietnam—Air Power." *Vital Speeches of the Day* 38, September 15, 1972.

—. "Strategy and Airpower: The Lessons of Southeast Asia." *Strategic Review* 1, Summer 1973.

Hopkins, Charles K. "Linebacker II: A Firsthand View." *Aerospace Historian* 23, September 1976.

Kamps, Charles Tustin. "Operation Linebacker II." *Air and Space Power Journal* 17, Fall 2003.

Leonard, Raymond W. "Learning from History: Linebacker II and U.S. Air Force Doctrine." *Journal of Military History* 58, April 1994.

"The Lessons of Vietnam; Exclusive Interview with Gen. Maxwell D. Taylor, USA (Ret), Adviser to Three Presidents." *U.S. News and World Report,* November 27, 1972.

"Nixon's Blitz Leads Back to the Table." *Time,* January 8, 1973.

Pape, Robert A. "Coercive Air Power in the Vietnam War." *International Security* 15, Fall 1990.

Parks, W. Hays. "Linebacker and the Law of War." *Air University Review* 34, January-February 1983.

Werrell, Kenneth P. "Linebacker II: The Decisive Use of Air Power?" *Air University Review* 38, January-March 1987.

"What the Christmas Bombing Did to North Vietnam." *U.S. News and World Report,* 18 February 5, 1973.

Wolff, Robert E. "Linebacker II: A Pilot's Perspective." *Air Force,* September 1979.

Yudkin, Richard A. "Vietnam: Policy, Strategy, and Airpower." *Air Force,* February 1973.